GW01312239

British Rail Main Line Electric Locomotives

The home depot for the Class 91s is Bounds Green on the ECML, from where most of the type test programme was undertaken. On the morning of 26th June 1988 a special line up of 'next generation' electric power was arranged at Bounds Green depot when Class 89 No. 89001, Class 90 No. 90005 and Class 91 No. 91002 were placed side by side.
Brian Morrison

A pair of Class 73/0 electro-diesel locomotives, Nos 73002 and 73003 approach Redhill on 30th August 1984 with the then daily Cliffe – Salfords aggregate train.

Colin J. Marsden

British Rail Main Line Electric Locomotives

Colin J. Marsden & Graham B. Fenn

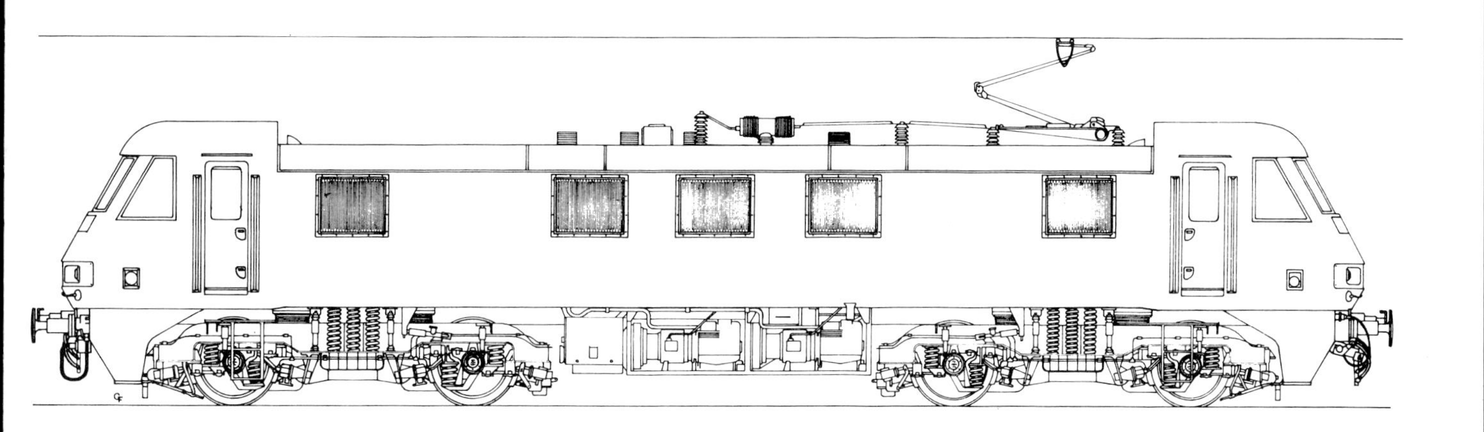

OPC

Oxford Publishing Co.

A catalogue record for this book is available from the British Library.

ISBN 0-86093-446-2

Oxford Publishing Co. is part of the
Haynes Publishing Group PLC
Sparkford, near Yeovil, Somerset, BA22 7JJ

Haynes Publication Inc.
861 Lawrence Drive, Newbury Park, California 91320, USA

Printed in Great Britain by Redwood Press Ltd

Note: Train times are as per BR timetables, therefore the 24-hour clock has been used for 1964 and onwards.

A line up of super power outside the Railway Technical Centre, Derby on 19th May 1988. Class 91 No. 91003, Class 90 No. 90008 and Class 89 No. 89001 pose outside the Engineering Development Unit while under preparation for transfer to Hamburg to take part in an International Rail Exhibition.

Colin J. Marsden

Title page: DWG 1 Class 90 'B' elevation. 3mm : 1ft scale.

Contents

Introduction

After completion of the title *British Rail Main Line Diesel Locomotives*, published in October 1988, the decision was taken by the publishers and authors to pursue a companion volume covering the main-line electric locomotive types and classes of BR and its constituent companies.

Again, it has been possible to produce a complete record in terms of text, technical information, photographic illustrations and 4mm:1ft scale drawings of all electric locomotive classes which are brought together for the first time in one book. For each type or class, roof, side and end elevations have been specially drawn by joint author Graham Fenn, for types or classes where significant detail differences exist or sub-classes have been formed, separate drawings have been produced for each variant where practical.

The choice of photographic illustrations has been made carefully to show as many modifications and details as possible, while still showing the machines in their working environment. At the beginning of each class or type section a full data panel has been included, this covering the major technical items such as dimensions weights and equipment types. In general both the railway supply industry and BR have been most helpful in supplying the required information and a special thank you must be given to the numerous industrial and railway officials.

All drawings reproduced in this book have been specially prepared – mostly from official company records, thus ensuring total accuracy.

We would like to thank the many people who have assisted with information and illustrations for inclusion in this book. It is hoped that readers of this major work will benefit from its contents and receive much pleasure from browsing through its pages.

Colin J. Marsden
Graham B. Fenn

ES1

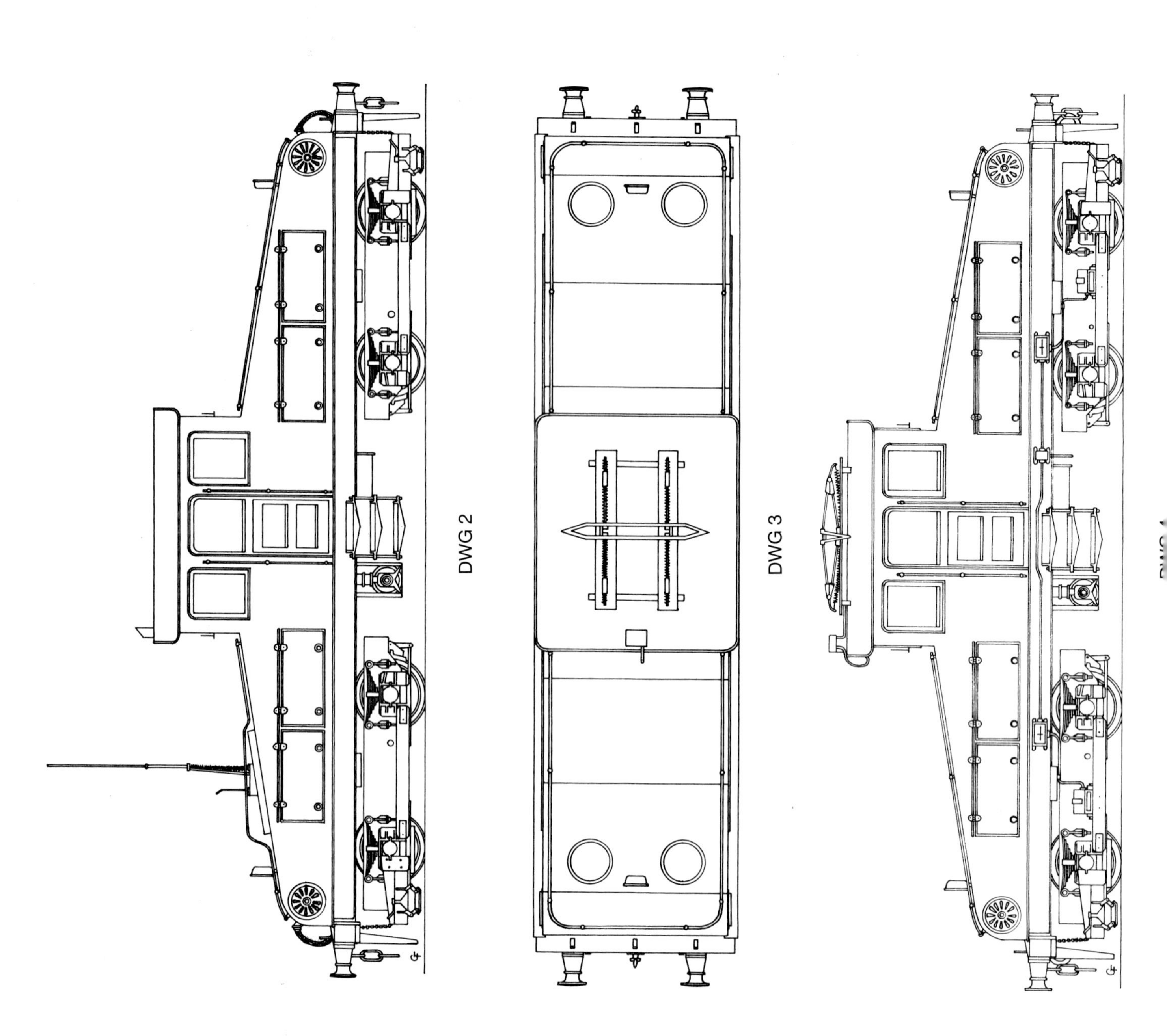

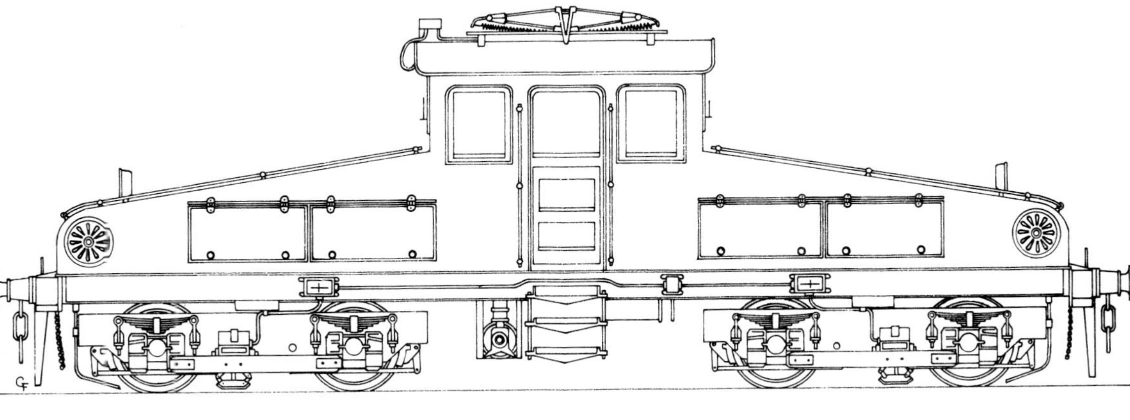
DWG 5

DWG 6

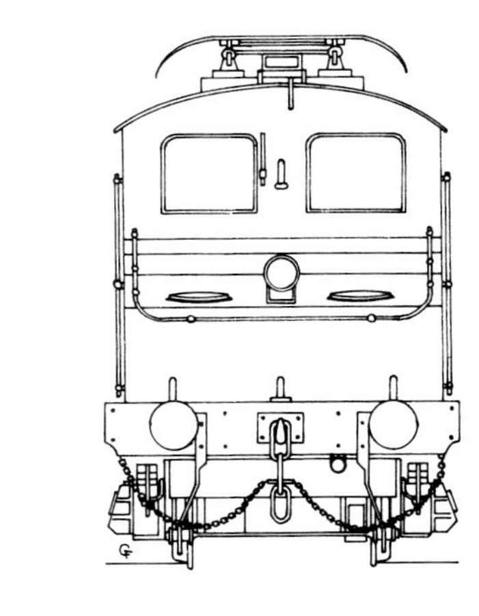
DWG 7

DWG 2
Class ES1 side elevation, showing original bonnet-mounted bow power collector.

DWG 3
Class ES1 roof layout, showing locomotive with roof-mounted power collector

DWG 4
Class ES1 side elevation showing locomotive with roof-mounted power collector and third rail pick-up slippers on bogie frame corners.

DWG 5
Class ES1 side elevation showing locomotive with roof-mounted power collector and third rail pick-up shoes mounted towards the centre of the bogie frame.

DWG 6
Class ES1 front end elevation, showing bonnet-mounted bow power collector.

DWG 7
Class ES1 front end elevation, showing roof-mounted power collector.

Class	ES1	Bogie Pivot Centres	20ft 6in
BR Number Range	26500-26501	Wheel Diameter	3ft
Former LNER Numbers	6480-6481	Brake Type	Air
Original NER Numbers	1-2	Sanding Equipment	Pneumatic
Built by	Brush	Heating Type	Non fitted
Introduced	1905	Coupling Restriction	Not multiple fitted
Wheel Arrangement	Bo-Bo 0-4+4-0	Horsepower	640 hp
Weight operational	56 tons	Tractive Effort maximum	25,000lb
Height - pan down	12ft 11in	Number of Traction Motors	4
Width	8ft 9in	Traction Motor Type	BTH
Length	37ft 11in	Control System	DC Direct
Maximum Speed	25mph	Gear Ratio	3.28:1
Wheelbase	27ft 8in	Nominal Supply Voltage	600V dc overhead and third rail
Bogie Wheelbase	6ft 6in		

The North Eastern Railway electrification scheme of the North Tyneside lines in 1904 included the one mile long freight only branch from Trafalgar Yard at Manors to Quayside Yard. The Quayside yard was around 130ft lower than the main line, on the banks of the River Tyne, access was arduous with a gradient of 1:27 including steep cuttings, and had a curved, single bore tunnel. Steam operation over the route caused appalling conditions, so the decision was taken to electrify the branch. Overhead power collection was the most desirable, but due to very limited clearances in the tunnel section, third rail power collection had to be used. However to avoid the presence of dangerous live rails in the yard areas, overhead power pick-up equipment was erected at these locations, which gave rise to dual power collection locomotives.

The North Eastern Railway gave authorisation to build two locomotives for the line, allocated numbers 1 and 2. These were centre cab locomotives with a bonnet at each end, and when built, a bonnet-mounted bow power collector for the overhead was fitted. For third rail power collection four standard slipper shoes, mounted on the corners of the bogies were fitted. Construction of the two locomotives was effected by Brush in 1905. Traction equipment was supplied by British Thomson-Houston (BTH).

At the Grouping, the NER was absorbed as part of the LNER, who renumbered these two electric locomotives as Nos 6480 and 6481. Under eventual BTC (BR) ownership in 1948 the locomotives were again renumbered to became Nos 26500 and 26501.

One noteworthy modification made to the locomotives was the replacement of the bow power collector by a cab roof-mounted cross-arm pantograph, which was done soon after entering service. At around the same time the bogie-mounted slipper pick-ups were moved from the ends to a more conventional mid-bogie position.

The two locomotives successfully operated the short line until 29th February 1964 when diesel traction was introduced, making these unique electric locomotives redundant. After their replacement the locomotives were stored at Gosforth and Heaton, being withdrawn the following September. Thankfully one of the pair, No. 26500, was saved from the breaker's yard and is now on display at the National Railway Museum, York.

Front 3/4 view of No.1, later Class ES1 No. E26500, showing the as-built condition with bonnet-mounted bow power collector, and third rail collector shoes mounted on the ends of the bogie frame. Note the omission of the third rail shoe fuse from the shoe beam, which was originally carried under the bufferbeam on the front of the locomotive.

Author's Collection

A general view of Class ES1 locomotive No. 26500 showing its later guise, with cab roof-mounted power collector, and third rail collector shoes mounted in the centre of the shoe beam. The livery applied is green adorned with both the second style BR crest, and a North Eastern Railway coat of arms. This view shows the locomotive at Heaton in August 1961.

L. Price

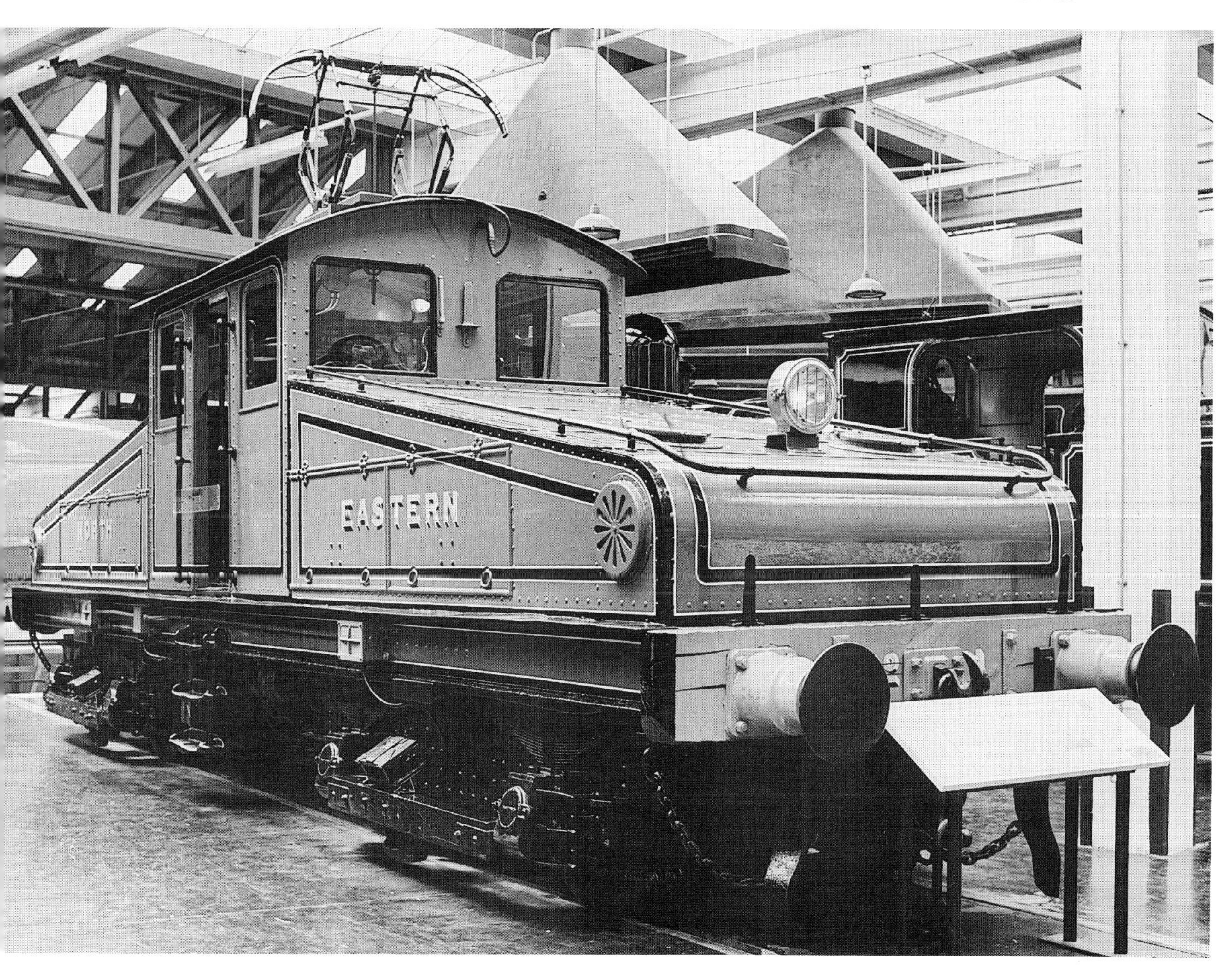

Thankfully one of this unique pair of electric locomotives has been saved, enabling today's enthusiasts to observe a Class ES1 at first hand. Locomotive No. 1, the later 26500, is preserved at the NRM York, where it is seen in Autumn 1988 painted in North Eastern Railway livery.

Colin J. Marsden

EB1

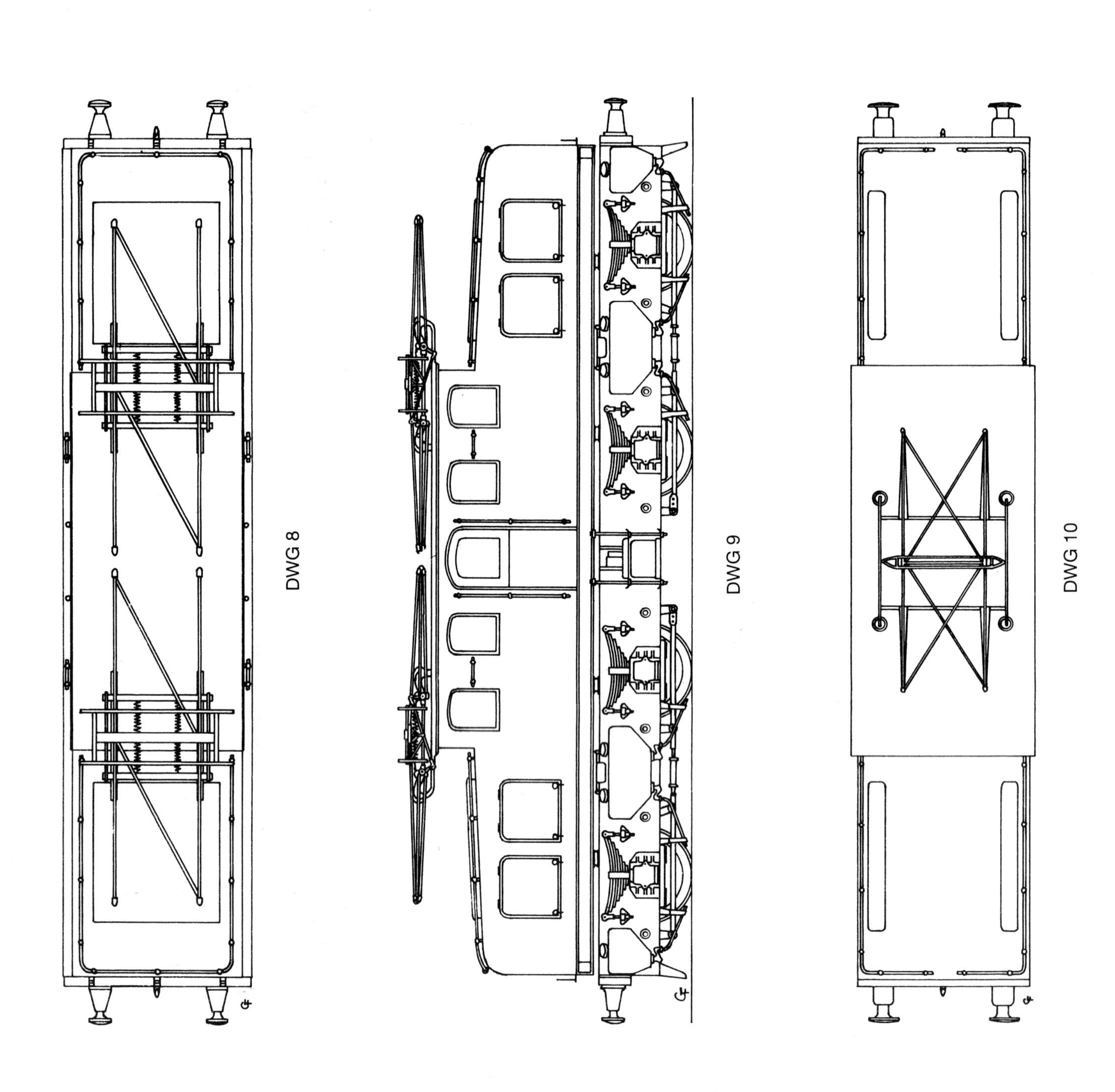

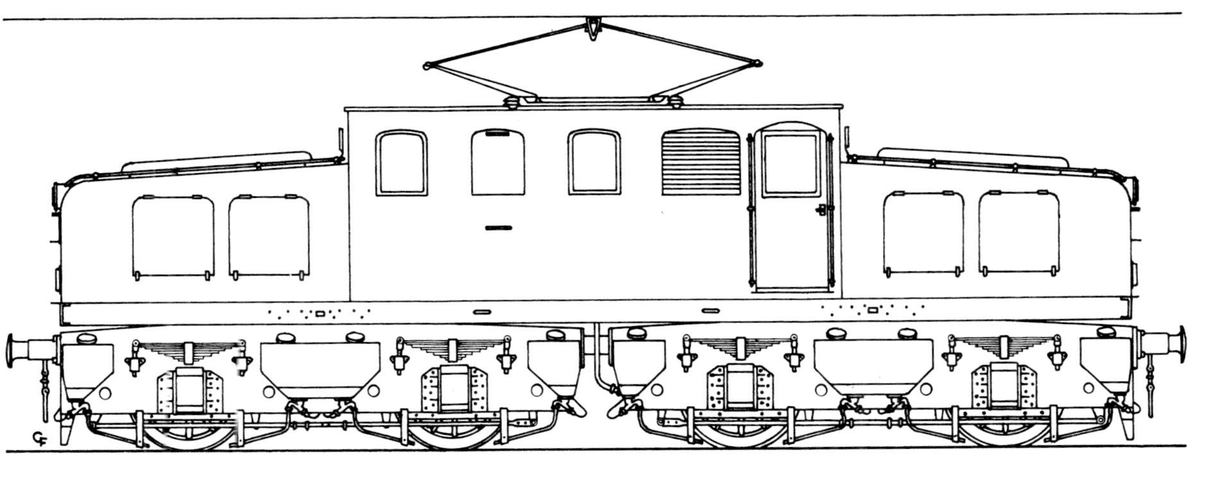

DWG 11

DWG 8
Class EB1 roof detail, showing original layout with two power collectors.

DWG 9
Class EB1 side elevation, showing original layout.

DWG 10
Roof detail of EB1 locomotive No. 11, modified for operation between Manchester and Sheffield.

DWG 11
Side elevation of EB1 locomotive No. 11, modified for operation between Manchester and Sheffield.

DWG 12
Class EB1 front end elevation.

DWG 13
Front end elevation of modified EB1 locomotive No. 11.

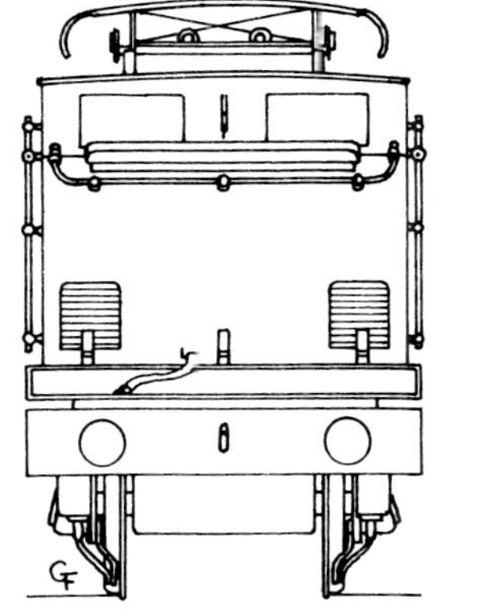

DWG 12

DWG 13

Class	EB1
BR Number Range	26502-26511
Former NER Number Range	3-12
Former LNER Number Range	6490-6499
Built by	NER Darlington
Introduced	1914
Wheel Arrangement	Bo-Bo (0-4+4-0)
Weight (operational)	75 tonnes
Height - pan down	13ft $1^{5}/_{16}$in
Width	8ft 4in
Length	39ft 4in
Maximum Speed	45mph
Wheelbase	27ft
Bogie Wheelbase	8ft 9in
Bogie Pivot Centres	18ft 3in
Wheel Diameter	4ft

Brake Type	Vacuum
Sanding Equipment	Pneumatic
Heating Type	Not fitted
Route Availability	Not issued
Coupling Restriction	Not multiple fitted
Horsepower	1,100hp
Tractive Effort (maximum)	28,000lb
Number of Traction Motors	4
Traction Motor Type	Siemens
Gear Ratio	4.5:1
Pantograph Type	Siemens
Nominal Supply Voltage	1,500V dc overhead

Locomotive No. 26510 was later rebuilt for the Manchester – Sheffield – Wath line, and subsequently taken into departmental stock as No 100.

The ten Bo-Bo Class EB1 locomotives were ordered by the NER in 1913 for use on the Newport – Shildon electrification scheme, authorised in the early months of the same year, in a vague attempt to save money by the use of electric traction. Under the NER the ten locomotives were allocated the numbers 3 - 12, while under subsequent LNER ownership the fleet became Nos 6490 - 6499, and eventually under the BR(BTC) banner Nos 26502 - 26511.

Construction of the locomotives was carried out at the NER Darlington workshops with Siemens Bros acting as chief sub-contractor supplying all electrical equipment. The building of the first eight locomotives was completed towards the end of 1914, but transfer to the electric system was not made until June 1915, when Nos 3-10 were sent to Shildon. Trial running commenced on 1st July 1915, and by early August an electrically operated freight service was introduced. The full compliment of locomotives was available for service from May 1920.

In daily operation the locomotives proved very reliable, but due to changing traffic flows, including the closure of Shildon marshalling yard, electric services ceased during January 1935, with the locomotives being transferred to Darlington for store. The fleet remained at Darlington until 1942, when No. 11 (LNER No. 6498) was moved to the LNER Doncaster Works for rebuilding as an experimental locomotive for use on the Manchester – Sheffield via Woodhead route, which was authorised for electrification in 1936. The rebuilding work carried out was considerable with the twin roof power collectors being replaced by a single pick-up, while body alterations included the repositioning of the cab doors. The output of the rebuilt locomotive was also amended from 1,100hp to 1,256hp and the tractive effort figure raised from 28,000lb to 37,000lb. Completion of the rebuilding work was in October 1944.

By 1947 the remaining nine locomotives at Darlington were moved to South Gosforth for storage, and prior to departure were renumbered into the LNER series. After being at South Gosforth for about a year, the locomotives were again renumbered into the BR series. Due to World War II the completion of the Manchester – Sheffield electrified system was deferred and in mid-1947 No. 6498, the former NER No. 11, was transferred from Doncaster Works to South Gosforth to join her sisters in store. In mid-1948 the locomotive was renumbered 26510.

Further life was found for No. 26510 in August 1949 when the locomotive was reallocated to Ilford electric depot on the ER(GE) system for carriage shunting. For its new role the locomotive was renumbered Departmental No. 100.

The nine locomotives stored at South Gosforth were eventually withdrawn on 21st August 1950 as surplus to requirements, and, with the exception of No. 26504, were sold to Wanty of Rotherham for scrap. No. 26504 was later dismantled by BR at Darlington to provide spare parts for the departmental survivor. The end of the road for the departmental locomotive came in 1960 when the 1,500V dc system of the GE was replaced by the standard 25kV ac system. The locomotive was dismantled by BR Doncaster Works.

The first of the Newport – Shildon 1,500V dc electric locomotives, No. 3 is seen posed in the works yard at Darlington on 11th May 1914, painted in workshop grey livery for the official photographs. The pantographs are raised, but there was of course no catenary available!

Author's Collection

Still painted in its original grey livery, No. 3 is seen near Simpasture on 16th July 1914 during one of its many trial runs before being permitted to enter normal service.

Author's Collection

NER No. 11 was the subject of a major rebuilding programme in 1942, when it was modified for use experimentally on the Manchester – Sheffield – Wath system. The modifications included the installation of one single roof mounted pantograph, the fitting of additional sandboxes, and a revised position for the cab door. After conversion the locomotive saw little use, but was later sent to Ilford for departmental operation in the depot where it was renumbered 100. As such it is seen at Ilford on 11th June 1960.

Alan Jackson

EE1

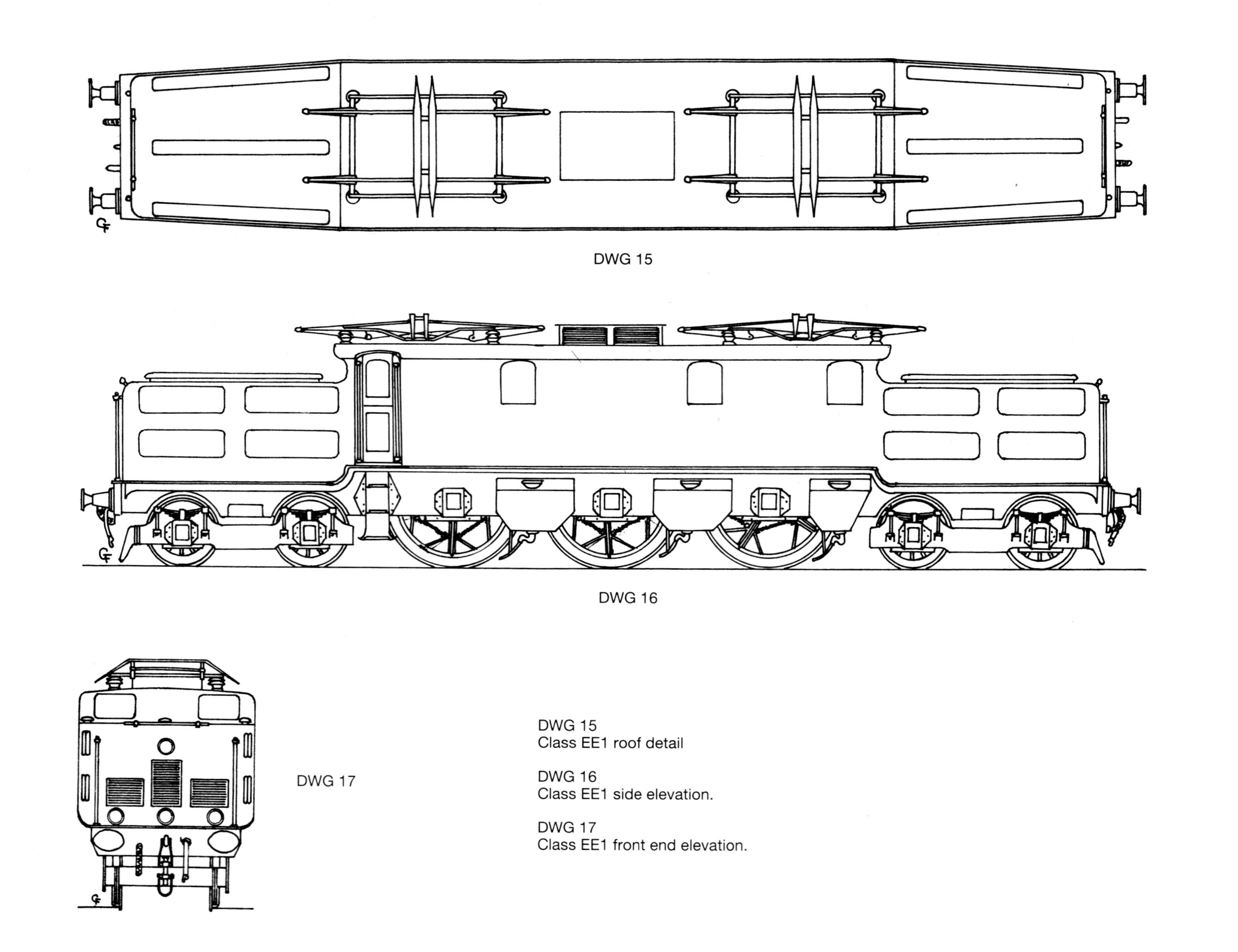

DWG 15
Class EE1 roof detail

DWG 16
Class EE1 side elevation.

DWG 17
Class EE1 front end elevation.

Class	EE1	Pivot Centres	37ft 2in
BR Number	26600	Wheel Diameter (Driving)	6ft 8in
Former BR Number	26699	Wheel Diameter (Pony)	3ft 7¼in
Former NER Number	13	Brake Type	Dual
Former LNER Number	6999	Sanding Equipment	Pneumatic
Built by	NER Darlington	Heating Type	Steam
Introduced	1922	Coupling Restriction	Not multiple fitted
Wheel Arrangement	2-Co-2	Horsepower	1,800hp
Weight (operational)	110 tons	Tractive Effort	28,000lb
Height - pan down	13ft 0⅛in	Number of Traction Motors	6
Width	8ft 10in	Traction Motor Type	MV
Length	53ft 6in	Gear Ratio	24:85
Maximum Speed	90mph	Pantograph Type	Siemens
Wheelbase	43ft 8in	Nominal Supply Voltage	1,500V dc overhead
Fixed Wheelbase	16ft	Boiler Water Capacity	570gal

During 1919 various memoranda and reports were prepared dealing with the projected electrification of the NER route between York and Darlington. After visits by NER engineering staff to the USA in 1920, to inspect their electric locomotives, the NER sought, and were granted, in March 1920, authorisation to construct a 'prototype' main-line electric locomotive, at a then projected cost of £20,000. Planning and design work was completed at the end of 1920 and a construction contract was placed with the NER works at Darlington in January 1921. Assembly of the locomotive progressed well and deliveries of the Metropolitan – Vickers power equipment arrived on time to permit delivery of the completed locomotive in March 1922. The NER class designation allocated was EE and the running number allocated was No. 13.

No. 13 was basically the same at both ends, with a lower bonnet section. However, the No. 1 end housed a train heating boiler. The central (slightly elevated) driving cab area accommodated two roof-mounted cross-arm power collectors. Driving positions were located on the right of the cab layout in either direction.

When built No.13 was outshopped from Darlington North Road Works in 'workshop grey'; it is unclear whether the locomotive ever acquired NER green livery, but by July 1925 No. 13 was painted in LNER green.

Following completion of the locomotive, the only section of line it could work over was between Newport and Shildon, where on 4th June 1922 No. 13 was the subject of extensive tests, hauling a rake of NER passenger stock. After these trials No. 13 was placed in store at Darlington, being exhibited at various rail events until 1935 from when it remained at Darlington Works until 1947. During 1947 No. 13 was renumbered to 6999 and transferred to South Gosforth, along with the Newport – Shildon locomotives. In 1948, after Nationalisation, No. 6999 was again renumbered to 26600.

It is, of course, sad to recall that electrification of the York – Newcastle section was not effected under this scheme and with no available or projected work foreseen the Raven designed prototype main line electric locomotive, No. 26600, was withdrawn in August 1950 and sold for scrap the following December to Wanty of Catcliffe, Rotherham.

Side elevation of No. 13, posed outside Darlington Works soon after completion in May 1922. The livery applied in this illustration is workshop grey, which was retained for the locomotive's original tests on the Newport – Shildon line.
GEC Traction

Class 70

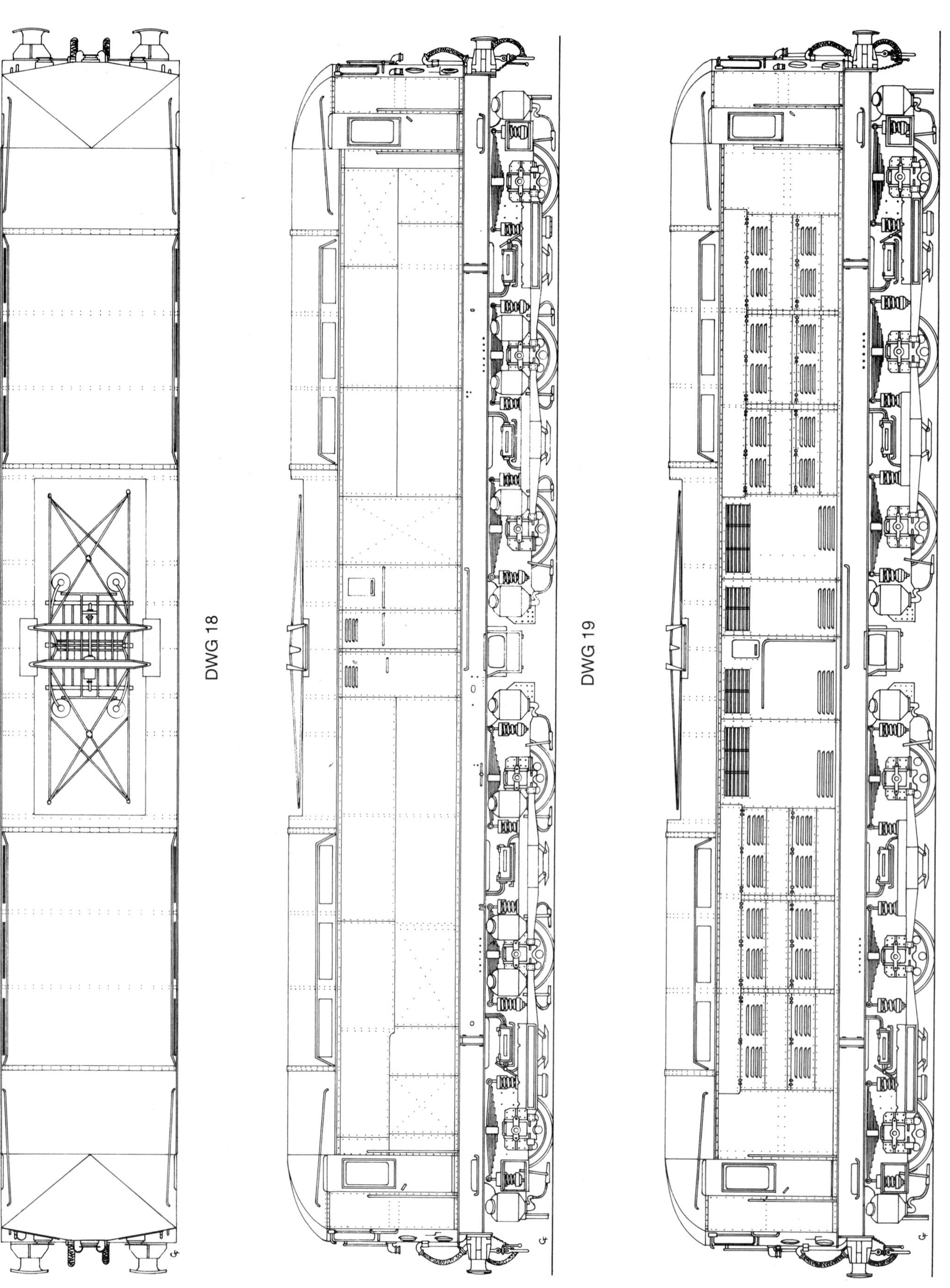

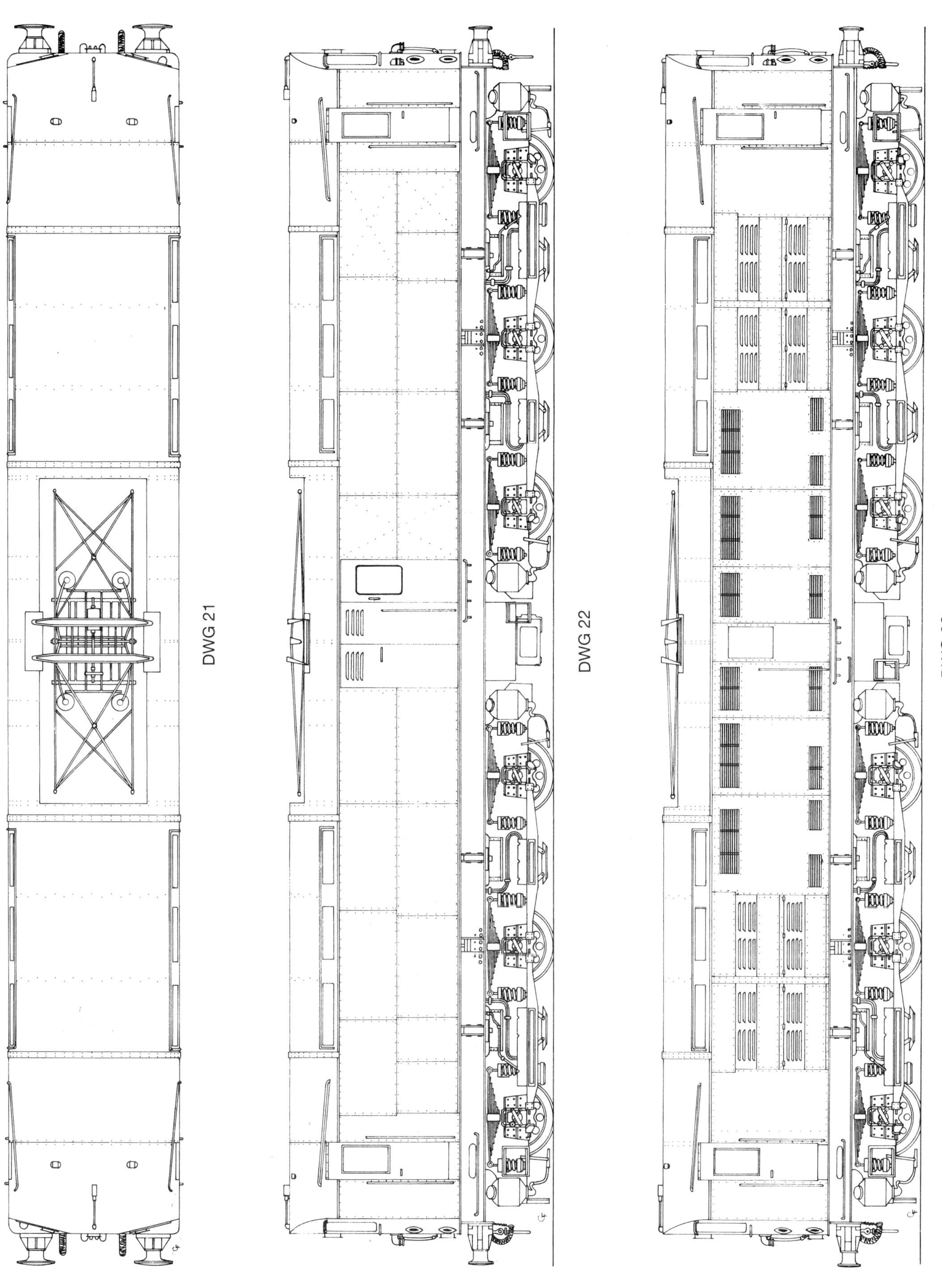
DWG 21
DWG 22
DWG 23

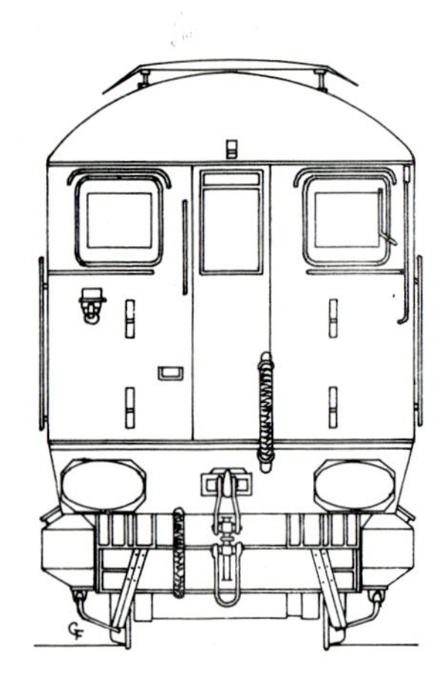

DWG 24

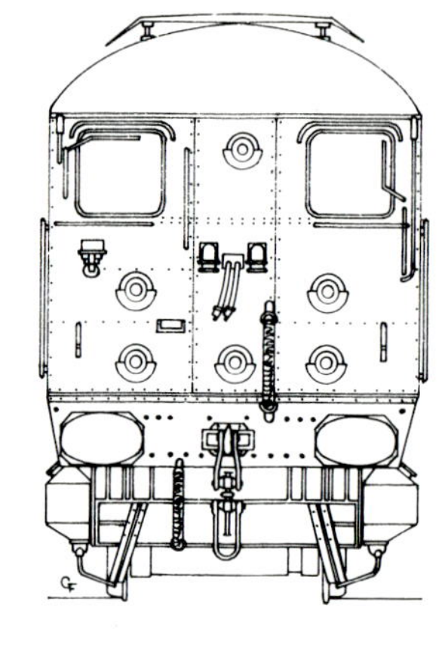

DWG 25

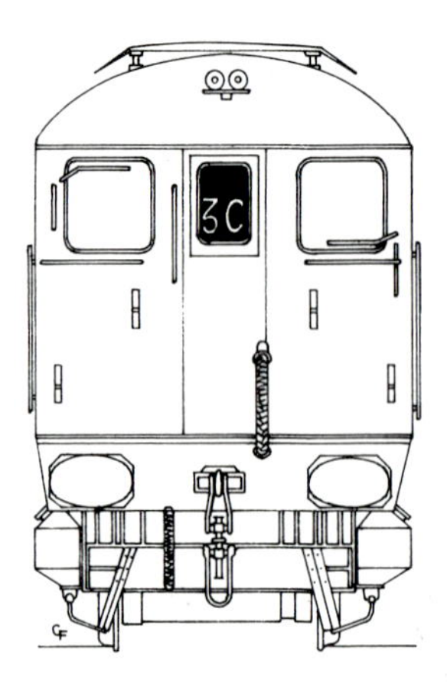

DWG 26

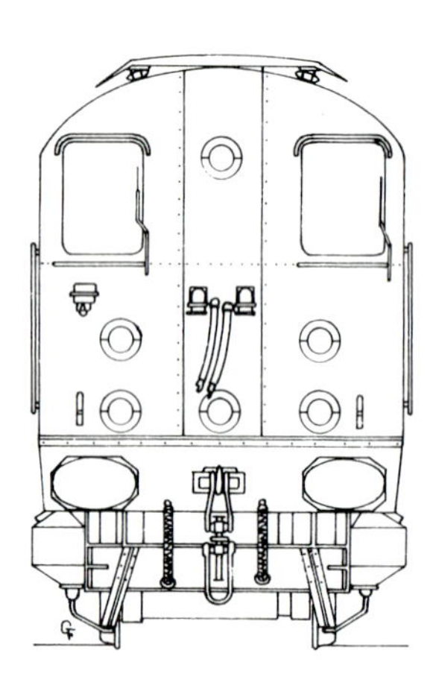

DWG 27

DWG 18
SR prototype electric locomotive roof detail, applicable to locomotives Nos CC1 and CC2 (20001 and 20002) in original condition.

DWG 19
SR prototype electric locomotive side 'A' elevation, as applicable to locomotives Nos CC1 and CC2 (20001 and 20002), showing original bogie style with five sand boxes per bogie.

DWG 20
SR prototype electric locomotive side 'B' elevation, as applicable to locomotives Nos CC1 and CC2 (20001 and 20002), showing modified bogie detail, ie with two sand boxes per bogie.

DWG 21
Roof detail of SR/BR electric locomotive No. 20003.

DWG 22
Side 'A' elevation of BR/SR electric locomotive No. 20003, showing as-built condition.

DWG 23
BR/SR electric locomotive No. 20003 side elevation of 'B' side, showing as-built condition.

DWG 24
Front end detail of SR prototype locomotives Nos CC1 and CC2 (20001 and 20002), showing original condition with stencil style headcode position.

DWG 25
Front end detail of SR prototype locomotives Nos CC1 and CC2 (20001 and 20002), showing the addition of multiple operation jumper equipment, waist height air connections and marker lights.

DWG 26
Front end detail of SR prototype locomotives Nos CC1 and CC2 (20001 and 20002), showing revised front end with two-character roller blind route indicator.

DWG 27
Front end detail of locomotive No. 20003, showing the original as-built style.

Class	70	70	70
Former SR Classification	CC	CC	
BR Number	20001	20002	20003
Southern Railway Number	CC1	CC2	Not Allocated
Built by	SR Ashford	SR Ashford	BR Brighton
Introduced	1941	1945	1948
Wheel Arrangement	Co-Co	Co-Co	Co-Co
Weight (operational)	100 tons	100 tons	105 tons
Height - pan down	12ft 6in	12ft 6in	12ft 8in
Width	8ft 7¼in	8ft 7¼in	8ft 7¼in
Length	56ft 9in	56ft 9in	58ft 6in
Min Curve negotiable	5½ chains	5½ chains	5½ chains
Maximum Speed	75mph	75mph	75mph
Wheelbase	43ft 6in	43ft 6in	44ft 6in
Bogie Wheelbase	16ft 0in	16ft 0in	16ft 0in
Bogie Pivot Centres	27ft 6in	27ft 6in	28ft 6in
Wheel Diameter	3ft 7in	3ft 7in	3ft 7in
Brake Type	Vacuum	Vacuum	Vacuum
Sanding Equipment	Pneumatic	Pneumatic	Pneumatic
Heating Type	Steam - Bastian & Allan	Steam - Bastian & Allan	Steam - Bastian & Allan
Boiler Water Capacity	320gal	320gal	320gal
Multi Coupling Restriction	Within type only	Within type only	Within type only
Brake Force	85 tons	85 tons	89 tons
Horsepower	1,470hp	1,470hp	1,470hp
Tractive Effort (maximum)	49,000lb	49,000lb	45,000lb
Number of Traction Motors	6	6	6
Traction Motor Type	EE 519A	EE 519A	EE 519-4D
Control System	DC Booster	DC Booster	DC Booster
Pantograph Type	EE Cross-arm	EE Cross-arm	EE Cross-arm
Nominal Supply Voltage	600-750V dc	600-750V dc	600-750V dc

Following the rapid extension of the third rail passenger network of the Southern Railway in the 1920s, it became apparent that the freight operation would benefit from electric propulsion. In the mid-1930s the Southern Railway Special Development Division at London Bridge produced plans for three prototype electric locomotives – these were originally intended to be of Bo-Bo design with an output of 1,500hp. The mechanical portion of the design became the responsibility of Southern Railway CME, R.E.L Maunsell, while the electrical equipment was designed by A. Raworth - Chief Electrical Engineer. By the time all designs and subsequent modifications had been put together the projected weight was around 84 tons, far too heavy for a Bo-Bo axle configuration, therefore a further re-design took place to produce a Co-Co locomotive design.

During the course of the Co-Co re-design, O.V.S Bulleid became the new Chief Mechanical Engineer of the SR, and took over the project. Construction of the first locomotive commencing at Ashford in 1940, and after completion it was taken to Brighton for technical fitting out. The second locomotive of the order followed, but was not completed until 1945. The third locomotive incorporating many technical and structural changes, did not appear on the main line until 1948.

During design work much attention was given to the bogies as the Co-Co configuration led to difficulty in providing a suitable bogie centre. The design adopted for this type had a large segmental bearing to improve ride quality. At an early stage it was intended to articulate the bogies and mount the drawgear on the end, however the final design was more conventional with the buffers and connections mounted on the body frame. When originally released to traffic the bogies were centrally attached, but the connector was later removed.

Power collection on this design was effected by two collector shoes on either side of each bogie, as well as a single cross-arm pantograph mounted centrally within the roof, the latter being used in yards where the presence of live rails would have been dangerous. In order to overcome the problem of 'gapping' due to the short shoebase a flywheel was installed which kept an electric 'booster set' running while short gaps were encountered.

The length of the locomotives was 56ft 9in for the original two and 58ft 6in for the third, while the width of all three was just over 8ft 7in, permitting use over the Hastings route, which was not electrified!

The first of the Co-Co locomotives, No. CC1, emerged in July 1941 and commenced trial running, mainly being deployed on freight traffic. primarily in its early days on wartime duties. No. CC2 emerged from Brighton Works in mid-1945 and was almost identical to No. CC1 except for technical modifications incorporated in the light of operating experience. The third locomotive of the design emerged after Nationalisation in 1948 and was numbered 20003 in the new BTC main line numeric sequence. This locomotive had a number of differences from its predecessors, it incorporated a larger flywheel, modified traction motors and modified body styling, with a more pleasing frontal design.

Although intended primarily for freight duties, by May 1949 all three were deployed on Victoria – Newhaven boat trains as well as Central Division freight services. The small fleet of three remained on the Central Division until 1968 when all were deemed as surplus to requirements and withdrawn.

When built Nos CC1 and CC2 were painted in SR green, while No. 20003 emerged in main line black/silver livery. In later years standard rail blue was applied to all three examples, and two-character headcodes replaced disc identification.

Southern electric locomotive No. CC1, the later 20001, is illustrated at the head of a freight train just prior to Nationalisation. Note that the former stencil route indicator position had been replaced by a marker light.

L. Price

Displaying its 'main line' black and silver livery, No. 20001 is seen inside Brighton Works during a major overhaul in the summer of 1960.

T. Wilson

During the mid-1950s the main line black and silver livery gave way to malachite green, offset by a white/red/white body stripe mid way up the body side. No. 20001 is seen passing Clapham Junction on 23rd September 1958 with the 9.30am Victoria – Newhaven Harbour boat train service.

John Faulkner

Sporting the 'British Railways' legend on its body side, the second of the SR electric locomotives, No. 20002 is illustrated painted in green livery in 1950.

T. Wilson

With the customary luggage vans coupled behind the locomotive, No. 20002 storms through Three Bridges on 12th April 1958 with an up Newhaven – Victoria boat train. The route diverging to the left by the locomotive was the line to East Grinstead.

John Faulkner

A route that became synonymous with the early SR electric locomotives was the Victoria – Newhaven line. Here, on 15th May 1949 the final member of the trio, No. 20003 is seen departing from Victoria with the first regular boat train service.

BR

It was rare to find the pantographs raised on the three SR, later Class 70, locomotives. However this illustration of No. 20003 with the pantograph raised, although devoid of an overhead supply, has been included to show the mechanical construction of the pick up assembly.

BR

Although ordered by the Southern Railway, the final locomotive of the class was delivered after Nationalisation, and thus assumed its BR number 20003 from new. Here we see No. 20003 brand new, painted in main line black and silver livery at Brighton.

BR

After only a relatively short period painted in main line black and silver livery, No. 20003 was repainted in a more suitable malachite green livery, offset by a white/red/white body side band. When painted in this livery the roof and sole bar were finished in grey.

J. Harris

In the later years the locomotives were modified to carry standard two-position headcode boxes fitted with roller blind units. No. 20003 painted in green livery shows this modification on 17th September 1966, while passing Tooting Bec with the 09.50 Victoria – Newhaven Harbour boat train.

John Scrace

Class 71

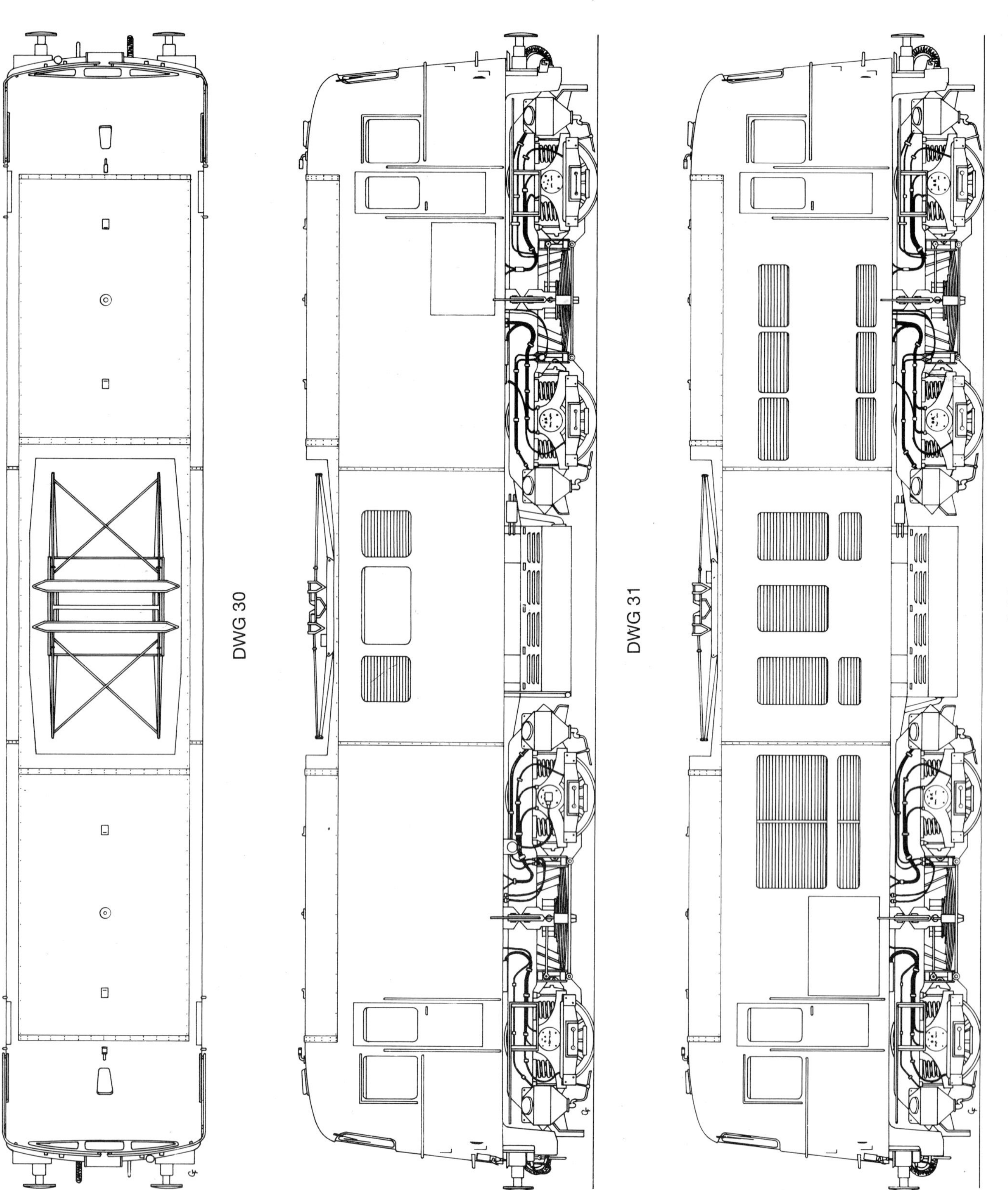

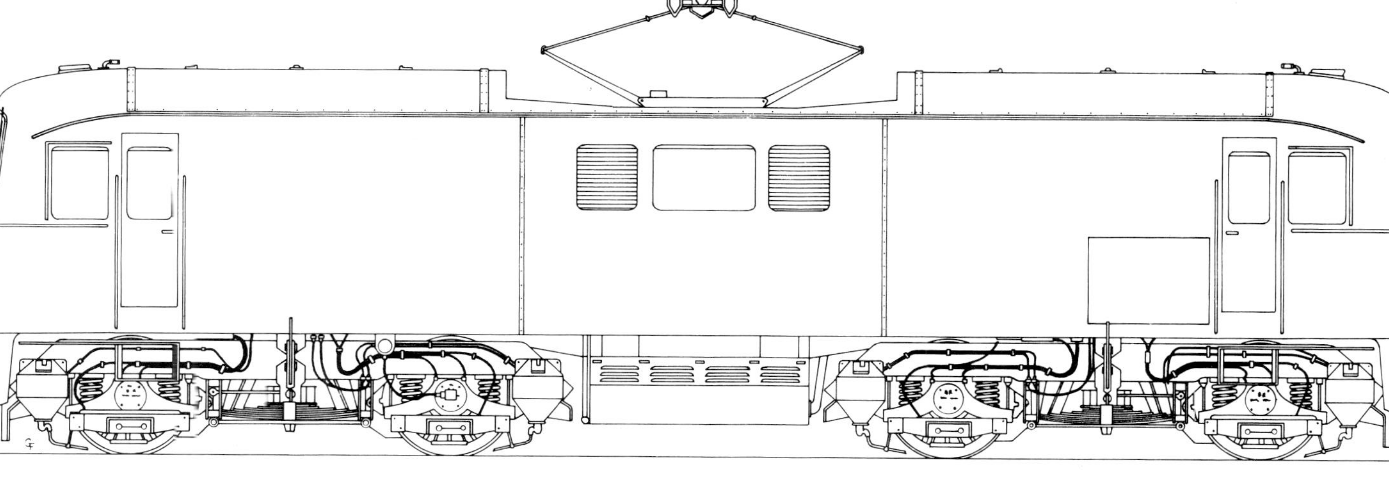

DWG 33

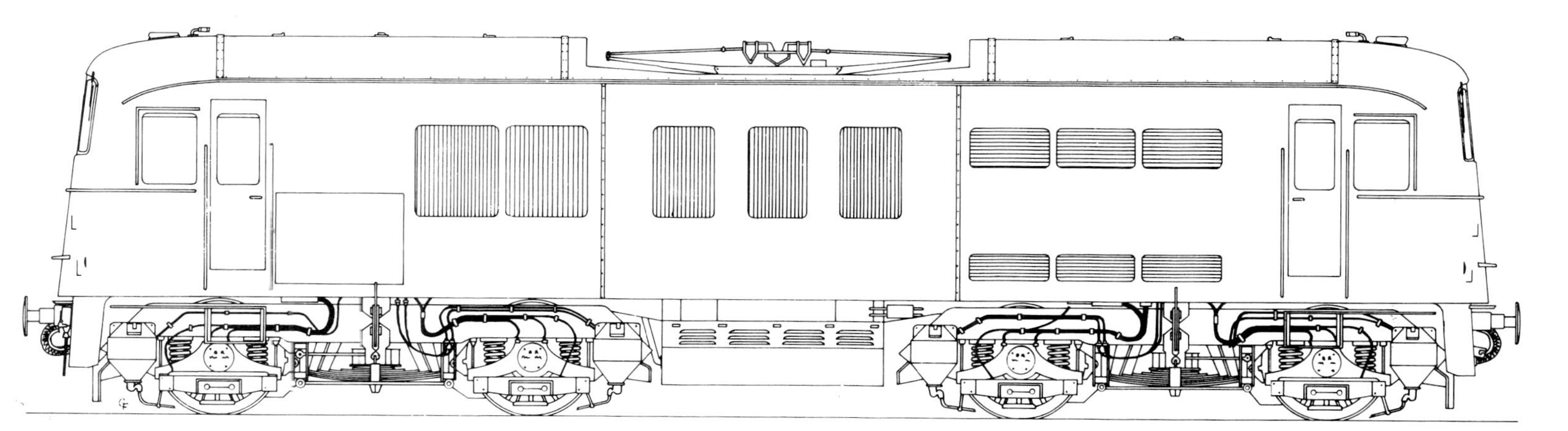

DWG 34

DWG 30
BR Bo-Bo electric (later Class 71) roof detail.

DWG 31
BR Bo-Bo electric (later Class 71) side 'A' elevation, showing original layout with body-mounted eth jumper, no rain water strip above cab windows and round sand box fillers.

DWG 32
BR Bo-Bo electric (later Class 71) side 'B' elevation, showing original layout, without rain water strip above cab windows and round sand box fillers.

DWG 33
Class 71 side 'A' elevation, showing pantograph in raised position, revised rain water strip above cab side windows, revised eth jumper position and square hinged sand box fillers.

DWG 34
Class 71 side 'B' elevation, showing modified side louvre arrangement applied to locomotives Nos E5004 and E5011 (71004 and 71011), revised rain water strip, eth jumper position and square sand box filler ports also shown.

DWG 35
BR Bo-Bo (later Class 71) front end layout, showing original condition.

DWG 36
Class 71 front end layout, showing revised style.

DWG 35

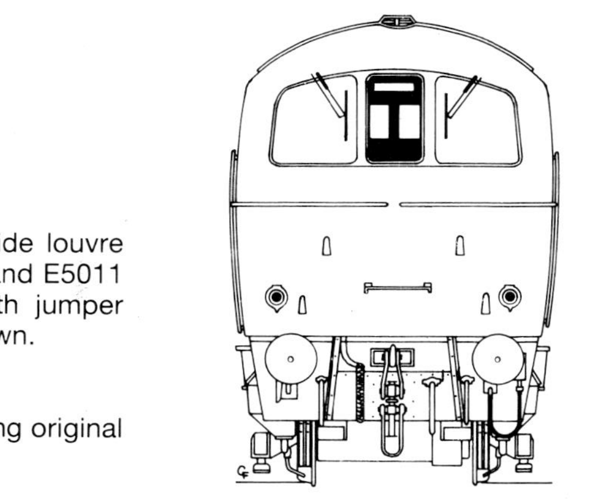

DWG 36

Class	71
Number Range (TOPS)	71001-71014
Former Number Range	E5000-E5024 (Note: 1)
Built by	BR Doncaster
Introduced	1959-60
Wheel Arrangement	Bo-Bo
Weight (operational)	77 tons
Height - pan down	13ft 1in
Width	8ft 11in
Length	50ft 7in
Min Curve negotiable	4 chains
Maximum Speed	90mph
Wheelbase	37ft 6in
Bogie Wheelbase	10ft 6in
Bogie Pivot Centres	27ft
Wheel Diameter	4ft
Brake Type	Dual
Sanding Equipment	Pneumatic
Heating Type	Electric

Route Availability		6
Coupling Restriction		Not multiple fitted
Brake Force		41 tonnes
Horsepower	(continuous)	2,552hp
	(maximum)	3,000hp
Tractive Effort	(maximum)	43,000lb
Number of Traction Motors		4
Traction Motor Type		EE 532A
Control System		DC Booster EE836
Auxiliary Generator		EE 910B
Gear Ratio		76:22
Pantograph Type		Cross-Arm
Nominal Supply Voltage		660-750V dc

Note: 1 A total of 24 locomotives were constructed, ten being converted to Class 74 electro-diesel locomotives in 1966-67. When built the first locomotive was numbered E5000, but was renumbered as E5024.

Much was learned from the building and operation of the three Southern Railway Co-Co locomotives, and when the Modernisation Scheme orders were placed in 1955 for 'next generation' electric traction a batch of 24 'booster' electric locomotives were ordered for use on the then newly electrified Kent Coast system. Their intended use was to be on main line passenger/freight services. The 'booster' electric locomotive order was placed with BR Workshops Doncaster, from where the first locomotive emerged in late 1958. The number range allocated was E5000-E5023, which under the BR TOPS renumbering system was amended to the 710xx series.

This fleet, unlike their prototypes, was mounted on a Bo-Bo bogie configuration, but retained the ability to collect power from both the third rail and overhead system, via a single roof-mounted pantograph, or bogie mounted shoes. Again, in common with their predecessors the locomotives incorporated a 'booster' system with a large flywheel to assist in power continuity over gaps in the live rail. A change from the prototype fleet was the use of fully spring-borne traction motors and a flexible drive system from traction motors to axles.

The delivery of the E5000 fleet was made initially to Ashford (Chart Leacon) where commissioning and testing was carried out. Driver training was effected throughout 1959-60 on the South Eastern Division main line, and only included drivers from the South Eastern Division depots. By the end of 1960 all 24 locomotives were available and had entered service. The livery applied was BR green with a grey roof, on the first locomotive a red mid height body band was applied between the cab doors. Over the ensuing years the fleet, always remaining allocated to the South Eastern section, was deployed on main line freight services as well as the prestigious "Night Ferry" and "Golden Arrow" passenger duties.

In 1965-66 changing traffic flows led to ten members of the fleet being deemed as redundant. At the same time the SR were looking into the possibility of introducing 'high power' electro-diesel (dual-power) locomotives on the South Western Divisions Waterloo – Bournemouth line. Therefore a major rebuilding scheme was authorised for ten members of the Class 71 fleet. The now depleted Class 71s soldiered on until 1977 when the few diagrams that still existed were taken over by Class 33 and 73 locomotives. The entire fleet became extinct from December 1977. For many months after withdrawal members of the fleet could be found laying at a number of locations, including Ashford, Hither Green and Stewarts Lane. It was hoped at one time that further work might be found for them, but alas this was not to be, and all examples except No. 71001 were sold for scrap and broken up. No. 71001, the former No. E5001 was saved by the National Railway Museum at York and restored to as near as possible original condition by BREL Doncaster.

Until mid-1992, No. E5001 was a major exhibit at the NRM, making a number of trips to BR depots for display at open days. Following display at Ashford in mid-1992, the NRM gave authority for Chart Leacon to technically restore the locomotive to main line standards. Its return to BR running was on 12th September 1992 when it powered a private charter from Waterloo to Bournemouth. In the future the locomotive will be usd on further specials.

The second of the Doncaster-built SR electric locomotives, No. E5001 stands outside Stewarts Lane Electric Depot in February 1959, soon after arrival from the builder's works. Note the original position and style of the electric train heating jumper and the larger sized train reporting numbers.

Author's Collection

Apart from being deployed on the Victoria – Dover Continental boat train services, the SR electrics, classified 'HA' by the Southern, were used on various South Eastern section freight and van trains. No. E5001 is seen passing Shortlands Junction in 1960 with a vans train for Bricklayers Arms.

GEC Traction

To provide maintenance facilities for the new electric locomotives, a purpose-built inspection and repair shed was constructed at Stewarts Lane, Battersea. This view of the new building, taken on 2nd May 1959, shows three 'HA' electric locomotives receiving attention. In the foreground the depot's wheel lathe can be seen.

Jim Oatway

At the head of an engineers train, No. E5003 is seen near Sevenoaks during the autumn of 1961. The train consist is somewhat unusual as its contains passenger stock, a goods brake van, a coal wagon and engineers' department stock.

Author's Collection

For much of 1959 the SR Traction Training School was undertaking driver training courses on the 'HA' locomotives, these involving both classroom and practical tuition. In June 1959 No. E5003 is seen near St Mary Cray with an up training special bound for Stewarts Lane.

The late Derek Cross

The duties for which the 'HA', later Class 71s, will be best remembered must surely be the "Golden Arrow" and "Night Ferry" trains between London and the Continent, which the locomotives operated between Victoria and Dover. Bearing the "Golden Arrow" headboard No. E5007 passes Petts Wood Junction on 1st October 1968 with the down service.
John Cooper-Smith

On 3rd August 1971 No.E5009 traverses the coastway between Dover and Folkestone with an up Continental mail train. The first two vehicles of the formation are Royal Mail tender and sorting coaches.
John Cooper-Smith

After restoration to working condition, No. E5001 made its first public run, as the first preserved main line electric locomotive to operate on BR tracks on 12th September 1992, when it powered the 09.00 Waterloo – Bournemouth enthusiasts' special. The train is seen near Totton, with Class 73/1 No. 73132 coupled behind to provide a train supply.
Colin J. Marsden

Class 73

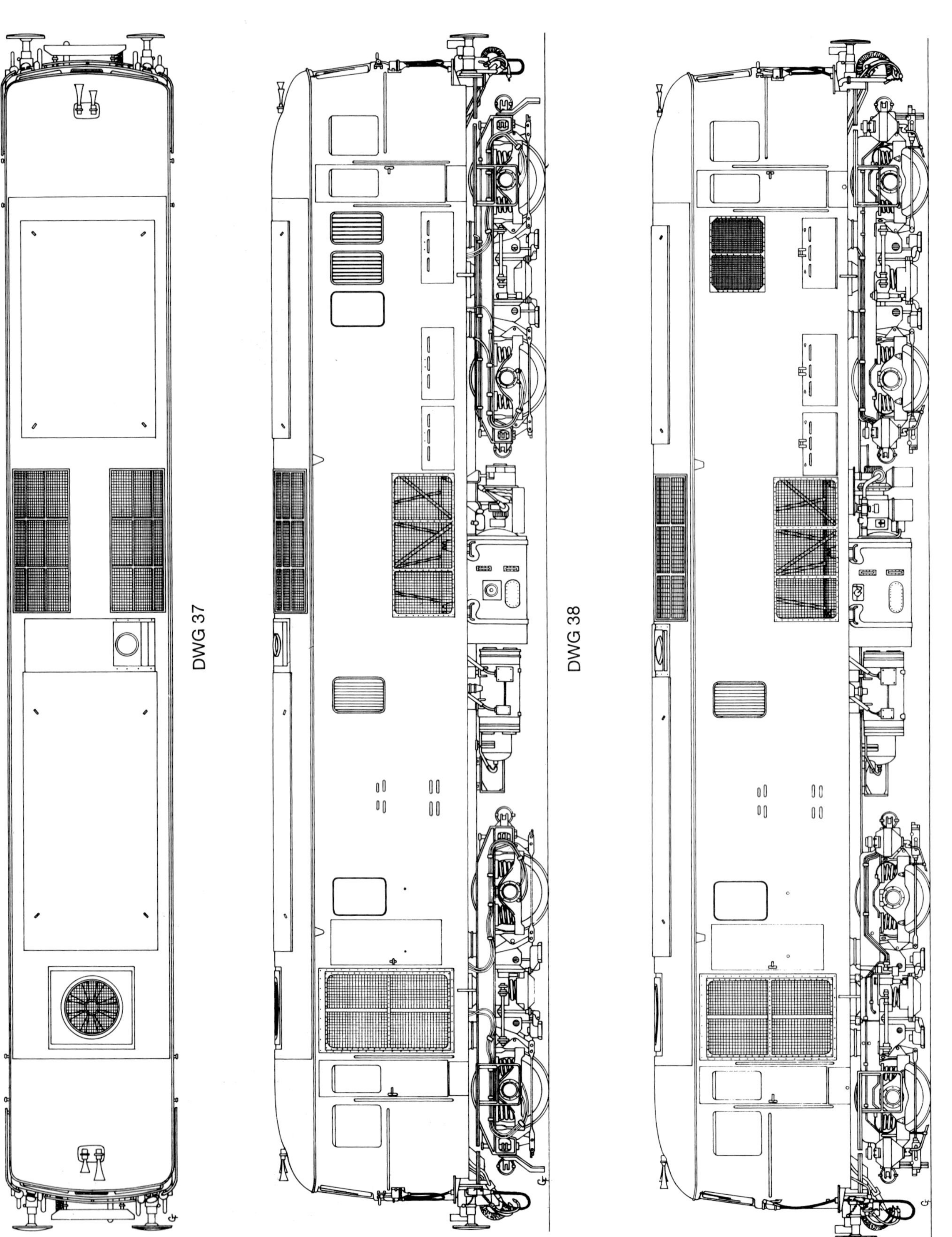

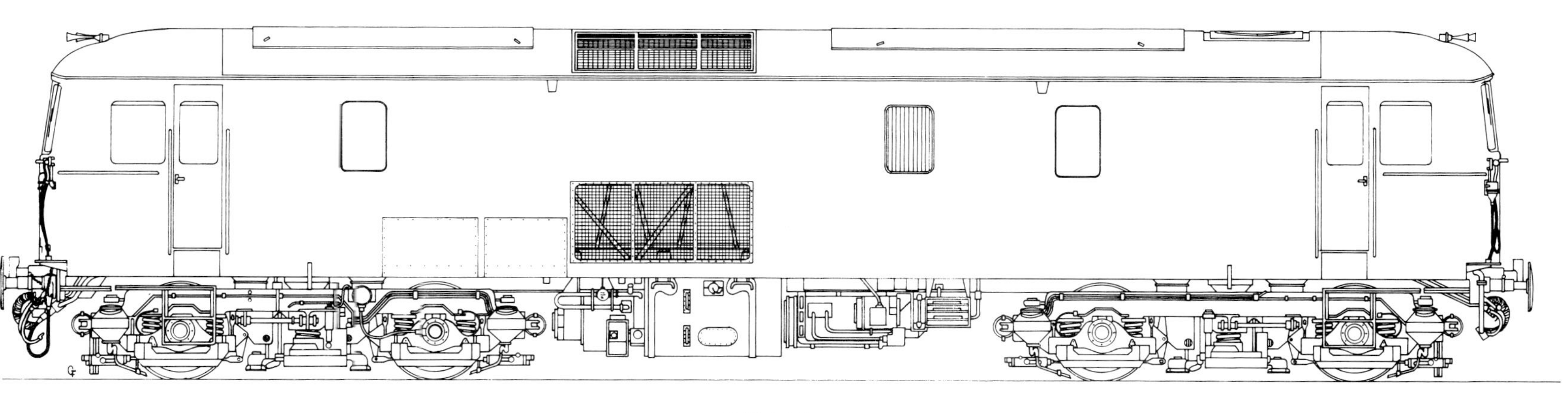

DWG 40

DWG 37
Class 73 roof detail, applicable to all sub classes, No. 1 end on left.

DWG 38
Class 73/0 'prototype' fleet (Nos E6001-E6006/73001-73006), side elevation. No. 1 end (diesel) on left.

DWG 39
Class 73/1 and 73/2 'production' fleet (Nos E6007-E6049/73101-73235), side elevation. No. 1 end (diesel) on left. Bogies show modified hinged lid sand boxes.

DWG 40
Class 73/1 and 73/2 'production' fleet (Nos E6007-E6049/73101-73235), side elevation. No. 1 end (diesel) on right. Bogies show original twist top sand boxes.

DWG 41
Class 73/0 'prototype' fleet (Nos E6001-E6007/73001-73006), front end, showing the later installed Oleo pneumatic buffers.

DWG 42
Class 73/1 and 73/2 'production' fleet (Nos E6007-E6049/73101-73235), front end layout.

DWG 42A
Class 73/1 and 73/2 'production' fleet (Nos E6007-E6049/73101-73235), front end layout showing sealed beam headlight.

DWG 41

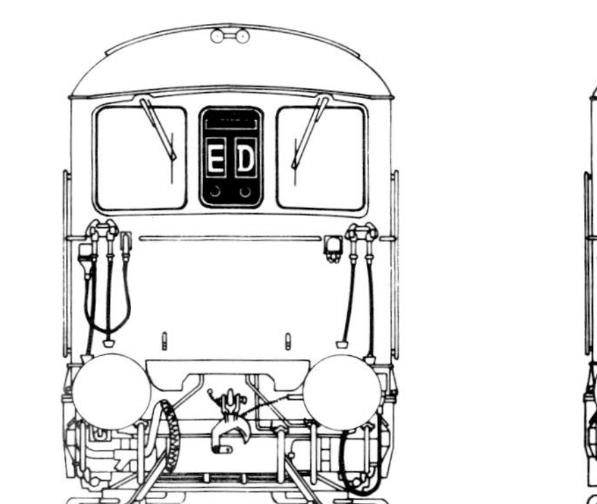

DWG 42

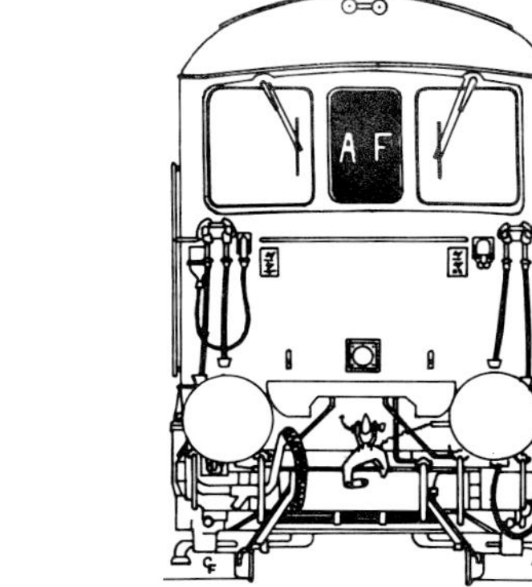

DWG 42A

Class	73/0	73/1	73/2
Former Class Codes	72	—	73/1
SR Class Code	JA	JB	JB
Number Range (TOPS)	73001-73006	73101-73142	73201-73235
Former Number Range	E6001-E6006	E6007-E6049	Random from 73/1
Built By	BR Eastleigh	EE Vulcan Foundry	EE Vulcan Foundry
Introduced	1962	1965-67	As 73/2 - 1988
Wheel Arrangement	Bo-Bo	Bo-Bo	Bo-Bo
Weight (operational)	76 tons	77 tons	77 tons
Height	12ft 5⁵/₁₆in	12ft 5⁵/₁₆in	12ft 5⁵/₁₆in
Width	8ft 8in	8ft 8in	8ft 8in
Length (Buffers extended)	53ft 8in	53ft 8in	53ft 8in
(Buffers retracted)	52ft 6in	52ft 6in	52ft 6in
Min Curve negotiable	4 chains	4 chains	4 chains
Maximum Speed	60mph (Note: 3)	60mph (Note: 3)	90mph
Wheelbase	40ft 9in	40ft 9in	40ft 9in
Bogie Wheelbase	8ft 9in	8ft 9in	8ft 9in
Bogie Pivot Centres	32ft	32ft	32ft
Wheel Diameter	3ft 4in	3ft 4in	3ft 4in
Brake Type	Dual, EP	Dual, EP	Dual, EP
Sanding Equipment	Pneumatic	Pneumatic	Pneumatic
Heating Type	Electric - Index 66 (Note: 1)	Electric - Index 66 (Note: 1)	Electric - Index 66 (Note: 1)
Route Availability	6	6	6
Coupling Restriction	Blue Star (Note: 2)	Blue Star (Note: 2)	Blue Star (Note: 2)
Brake Force	31 tons	31 tons	31 tons
Nominal Supply Voltage	600-750V dc	600-750V dc	600-750V dc
Engine Type	English Electric 4SRKT Mk 11	English Electric 4SRKT Mk 11	English Electric 4 SRKT Mk 11
Horsepower (Electric)	1,420hp	1,420hp	1,420hp
(Diesel)	600hp	600hp	600hp
Tractive Effort (Electric)	42,000lb	40,000lb	40,000lb
(Diesel)	34,100lb	36,000lb	36,000lb
Cylinder Bore	10in	10in	10in
Cylinder Stroke	12in	12in	12in
Main Generator Type	EE824-3D	EE824-5D	EE824-5D
Aux Generator Type	EE908-3C	EE908-5C	EE908-5C
Traction Motor Type	EE542A	EE546-1B	EE546-1B
Gear Ratio	63:17	61:19	61:19
Fuel Tank Capacity	340gal	310gal	310gal

Note: 1 ETS is only available under electric conditions. A pre-heat system is available on the Class 73/0s.

Note: 2 Multiple coupling conforms to Blue Star for diesel operation, in addition each sub-class can operate in multiple together, and in multiple with Class 33/1 locomotives, as well as selected post 1951 emu stock using the 27 wire waist height jumper connections.

Note: 3 When built class 73/0 max speed was 80mph, and Class 73/1 90mph, now reduced to 60mph.

Class 73 Sub class differences:

Class 73/0 Prototype Electro-Diesel locomotive fleet, constructed by BR at Eastleigh, design proved suitable for SR operation, and production fleet ordered. The 73/0 fleet have an additional jumper cable on the nose end, and grille differences on body sides.

Class 73/1 Production fleet of Electro-Diesel locomotives, constructed by English Electric, mounted on revised bogie design, and have side grille differences to the Class 73/0.

Class 73/2 InterCity Gatwick locomotives.

For the Southern Region, where the principle power source was electric, a dual-power (electric and diesel) locomotive was considered a distinct advantage, having the ability to operate from the electric supply, or if this was unavailable, an on board subsidiary diesel could be started to provide power to the traction motors. The basic plans for this style of dual power locomotive were first considered in the late 1930s, but it was not until the mid-1950s that any firm plans on the dual power concept were drawn up.

Plans advanced in the closing years of the decade and by July 1959 approval was given to construct six prototype dual-power locomotives. The prime power source was simple – straight electric. The auxiliary source adopted was an English Electric 4SRKT engine set to deliver 600hp, traction power being provided by an English Electric generator group. The electric power output was 1,600hp.

The construction contract for the six prototype locomotives was awarded to Eastleigh Carriage Works, where construction commenced in 1960. The first completed locomotive emerged on 1st February 1962 carrying the number E6001. The livery applied was BR green with small yellow warning panels. By the end of 1962 all six locomotives were in service, allocated to Stewarts Lane depot in South London and performing very well. An important design feature of the electro-diesel fleet was the installation of electric and diesel multiple control equipment as well as emu compatible jumpers, permitting the locomotives to operate in multiple with any blue star compatible locomotive or post-1951 electric multiple unit.

During the mid-1960s the SR became so pleased with their new charges that repeat orders for 43 almost identical locomotives were made in 1964, although the contract went to English Electric rather than BR. The number range allocated to these locomotives was E6007-E6049. The English Electric contract was carried out by the Vulcan Foundry Works at Newton-le-Willows, with the first locomotive arriving on SR metals in October 1965. The body styling of the EE product was practically identical to the BR build except for slight window/louvre alterations, the removal of the multiple control jumper from the driver's side and re-designed bogies. Minor internal technical alterations were also incorporated.

Under the BR numerical classification the fleet became Class 73. Sub Class 73/0 – Nos 73001-73006 for BR built locomotives and Sub Class 73/1 – Nos 73101-73142 for the EE built locomotives.

After their full introduction, the fleet of 49 locomotives settled down well, operating on all three SR divisions at the head of both passenger and freight services. The fleet's duties remained almost the same until May 1984 when the "Gatwick Express" service was launched. This called for seven of the fleet to be dedicated to the service daily.

For the "Gatwick Express" service special rakes of MkII stock were formed with a Gatwick Luggage Van coupled at one end, and a Class 73/1 at the other. Additional work was found for the fleet in 1986-88 when the new Bournemouth stock was under construction and the REP emus were phased out early to donate their electrical equipment to the new type. As a temporary measure Class 73s were formed with TC stock and used on some Waterloo – Bournemouth services.

When built the six prototypes were finished in BR green livery, however when the production EE fleet emerged, Electric blue was applied, this giving way to standard Rail blue from the late 1960s. The adoption of the 'more yellow' scheme for some classes was extended to the Class 73 fleet in 1983 but after only a handful had been so treated full main-line or InterCity colours were authorised, this being applied to all locomotives from 1984 to 1988. Following the introduction of the independently funded business groups within BR, other liveries have been applied. Some of the Class 73/1 members have appeared in Civil Engineers 'Dutch' livery, while others have been repainted in all over grey. From 1991 Network SouthEast livery was applied to locomotives owned by this business. Most members of the Class 73/0 fleet have retained the all-blue or 'more yellow' format. Following the re-introduction of names on selected fleets the Class 73s have benefited from this addition.

Following the take over of the "Gatwick Express" services it was decided in 1988 to dedicate a fleet of twelve locomotives specifically for this service, and classify them as Class 73/2. These locomotives, apart from sporting full InterCity colours are maintained for full 90mph running, whereas all others are now restricted to 60mph.

Over the years few modifications have been necessary to the Class 73 fleet. Two worthy of mention are (1) when built the six prototype machines had oval buffers which were later changed to the conventional Oleo type, and (2) following introduction on the "Gatwick Express" service several electrical fires occurred due to arcing between shoes and bogies, and to overcome this arc shields were fitted.

During 1991, it was agreed to repaint No. 73101 in full 'Pullman' livery. This was carried out by Selhurst Level 5 depot, and the locomotive was used to operate the VSOE on a special London – Brighton run. It was the intention to repaint the locomotive back into conventional livery, but agreement was reached to retain this distinctive livery. Following the decision to fit headlights to all main line traction from 1990, most members of the class have now been fitted.

The first of the SR dual power electro-diesel locomotives emerged from Eastleigh Carriage Works in January 1962. This view, taken on 28th January, shows No. E6001 parked in Eastleigh Works yard before operating a test special under diesel conditions to Basingstoke.

BR

When the production batch of electro-diesels emerged from English Electric a blue livery off-set by a grey base band was adopted in place of the original green. No. E6010 is seen alongside 'West Country' class 4-6-2 No. 34023 *Blackmore Vale* at Waterloo on 16th October 1966.

Colin J. Marsden

At the time of introduction of the electro-diesels the new standard BR rail blue livery had not been instigated and a number of livery variants were recorded. No. E6018 was no exception, painted with full wrap-round yellow ends and a BR electric blue body. The locomotive was photographed heading a rake of TC stock at St Denys on 17th June 1967 forming the 08.45 Waterloo – Lymington Pier service.

John H. Bird

The Class 73 fleet of electro-diesels have operated over the entire area of the former Southern Region and have indeed visited many non-electrified areas of the Southern and other regions, making the Class 73s some of the most versatile locomotives on BR. All-blue liveried No. 73006 is seen near Weybridge on 9th July 1985 with an empty vans train from Southampton bound for Clapham Junction yard.

Colin J. Marsden

Following the introduction of the "Gatwick Express" from May 1984, at first a batch of Class 73/1s were dedicated to this duty. All-blue No. 73109 hurries past Merstham on 26th April 1986 with the 09.00 Victoria – Gatwick service, consisting of a five-vehicle formation of Class 488 stock and a GLV on the rear.

Colin J. Marsden

When new or revised liveries were introduced in the mid-1980s the Class 73s were some of the first to benefit, with many examples repainted with wrap-round yellow ends, large logo and numbers and grey roofs. This livery was later superseded by the InterCity colour scheme. Wrap round yellow liveried No. 73131 passes Newnham on 23rd April 1987 with a vans train bound for Clapham Junction.

Colin J. Marsden

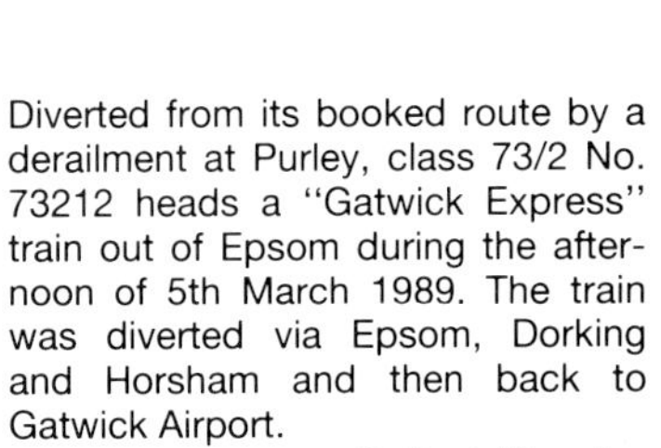

Diverted from its booked route by a derailment at Purley, class 73/2 No. 73212 heads a "Gatwick Express" train out of Epsom during the afternoon of 5th March 1989. The train was diverted via Epsom, Dorking and Horsham and then back to Gatwick Airport.

Colin J. Marsden

With the forming of BR into various operating sectors, a number of different liveries have been introduced. This has included InterCity, Main-Line, 'Dutch' and NSE. No. 73127 in InterCity /Main Line livery approaches West Byfleet on 10th May 1986 with a cement train bound for Northfleet.

Colin J. Marsden

Now classified as Class 73/2, No. 73201, the former No. 73142, has for many years been the SR's Royal Train locomotive, always being kept in an immaculate condition by Stewarts Lane depot. With a fresh coat of paint No. 73142 *Broadlands* approaches New Malden on 30th July 1986 with the 16.15 Waterloo – Southampton Eastern Docks Royal Train.

Colin J. Marsden

In the years 1986-88 the Waterloo – Bournemouth line went through a period of transition with the REP/TC stock gradually being replaced by new Class 442 'Wessex Electric' units. Whilst equipment was salvaged from the REP stock a number of trains were operated by Class 73/1s hauling TC vehicles. On 3rd July 1987 InterCity liveried No. 73107 passes Wimbledon West with the 10.00 Bournemouth – Waterloo service.

Colin J. Marsden

Class 73 locomotives owned by the Civil Engineers function were painted into grey/yellow or 'Dutch' livery in the period 1989-92, however following a further reorganisation of the businesses which took the engineering operations under the main operation businesses this livery may be short lived. Sporting 'Dutch' livery, No. 73128 *O.V.S. Bulleid CBE* is seen at Selhurst following repaint in September 1991.

Colin J. Marsden

One of the most surprising repaints to any locomotive during the 1990s, was the application of full 'Pullman' livery to No. 73101 *Brighton Evening Argus* in September 1991 by Selhurst Level 5 depot, the 'Pullman' insignia, motifs and numbers being supplied by the VSOE Company. The locomotive is seen here in Selhurst yard after completion.

Colin J. Marsden

From the commencement of the new Summer 1987 timetable all SR Western Section vans trains, hitherto maintained at Clapham Yard, were transferred to Eastleigh, thus eliminating the regular sight of Class 73s on van trains from the South Western Division main line during daylight hours. On 10th April 1987 InterCity liveried No. 73118 passes Woking Junction with the 09.25 Eastleigh – Clapham Junction. Since this picture was taken van trains have stopped running altogether on the former Southern Region.

Colin J. Marsden

Class 74

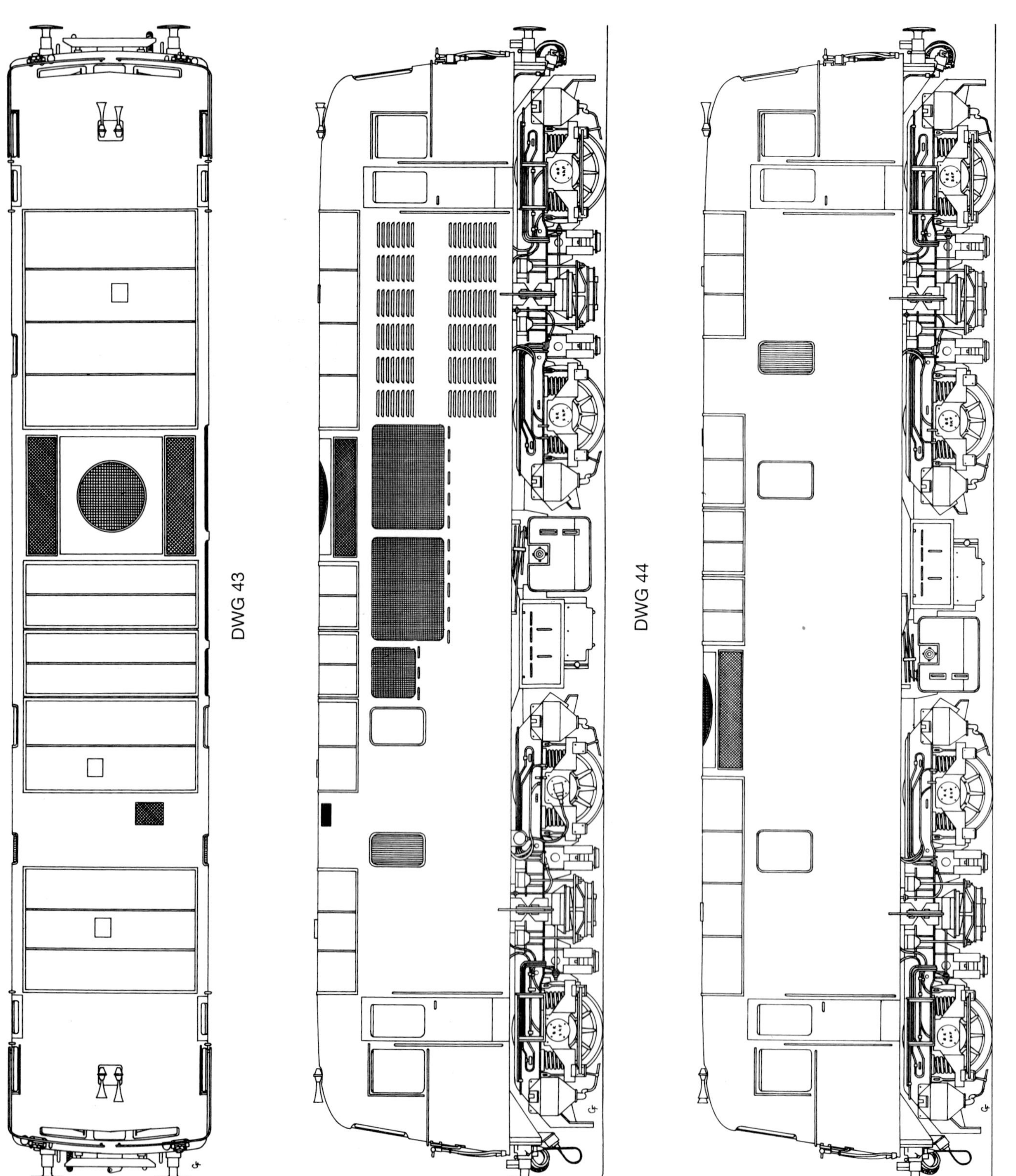

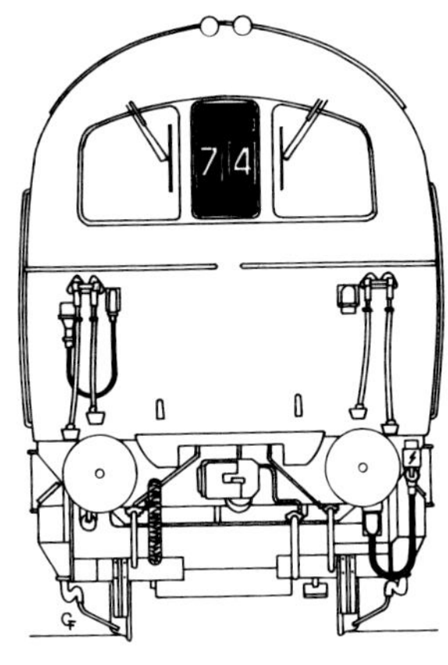

DWG 46

DWG 43
Class 74 roof detail, No. 1 end on right.

DWG 44
Class 74 side elevation, No. 1 end on right.

DWG 45
Class 74 side elevation, No. 1 end on left.

DWG 46
Class 74 front end layout.

Class		74
Number Range (TOPS)		74001-74010
Former Number Range		E6101-E6110 (Note: 1)
Rebuilt by		BR Crewe
Introduced		1966-68
Wheel Arrangement		Bo-Bo
Weight operational		86 tons
Height		12ft $9\frac{5}{8}$in
Width		9ft
Length	(Buffers extended)	50ft $5\frac{3}{4}$in
	(Buffers retracted)	49ft $3\frac{3}{4}$in
Min Curve negotiable		4 chains
Maximum Speed		90mph
Wheelbase		37ft 6in
Bogie Wheelbase		10ft 6in
Bogie Pivot Centres		27ft
Wheel Diameter		4ft
Brake Type		Dual
Sanding Equipment		Pneumatic
Heating Type		Electric - Index 66
Route Availability		7
Coupling Restriction		Blue star (Note: 2)

Brake Force		41 tons
Nominal Supply Voltage		600-750V dc
Engine Type		Paxman 6YJXL
Horsepower	(Electric)	2,552hp
	(Diesel)	650hp
Rail Horsepower	(Electric)	2,020hp
	(Diesel)	315hp
Tractive Effort	(Electric)	47,500lb
	(Diesel)	40,000lb
Cylinder Bore		7in
Cylinder Stroke		$7\frac{3}{4}$in
Main Generator Type		EE843
Number of Traction Motors		4
Traction Motor Type		EE532A
Gear Ratio		76:22
Fuel Tank Capacity		310gal

Note: 1 The Class 74s were converted from Class 71 locomotives Nos E5015/16/06/24/19/23/03/05/17/21.

Note: 2 Blue star multiple control equipment fitted for diesel operation. Locomotives also able to operate in multiple (electric) with Class 73, and most post 1951 electric multiple units.

With the coming of the Bournemouth electrification scheme in 1967 the Southern Region (SR) were anxious to exploit to the maximum the use of electric propulsion – however problems prevailed in some locations, such as Southampton Docks and between Branksome and Weymouth where finance was not available for full electrification. The SR with previous experience of dual power traction considered that a powerful electro-diesel would be the answer. At the same time, ten of the Region's E5000 type Class 71 electric locomotives were redundant; after much discussion between the SR, English Electric and the BR Workshops Division a major rebuilding programme for ten straight electric locomotives into 2,552/650hp electro-diesels was authorised. The rebuilding work was carried out at BR Crewe, with English Electric acting as chief sub-contractor. Following rebuilding the fleet was classified Class 74 (SR type HB), and renumbered in the range E6101-E6110. Under TOPS renumbering the fleet became Nos 74001-74010.

The rebuilding of this class was a major undertaking, which called for the entire body to be dismantled down to the frames and then rebuilt. For their new role the main electrical fittings including the booster set were retained, the auxiliary power source being provided by a new Paxman 6YJXL engine of 650hp.

Conversion work at Crewe was a protracted affair with the first machine not arriving on the SR until the end of 1967. All ten locomotives had reached the Region by June 1968. From delivery all locomotives were allocated to Eastleigh depot, and were always used on the South Western Division, where regrettably their performance was not always good, with frequent failures, mainly attributable to the large amount of sophisticated electronics installed. In common with other SR locomotive traction, multiple control jumper cabling was installed, enabling the locomotive to be operated with 1951, 1957, 1963 and 1966 type SR emu stock under electric conditions and blue star restricted locomotives under diesel conditions.

From new the Class 74s were painted in conventional Rail blue livery with full yellow warning ends, the only significant change being the alteration of the number sequence during the early 1970s when TOPS renumbering was introduced.

With the changing traction requirement of the SR and the fleet's record of misbehaviour, the Class 74s were doomed, with the entire fleet being withdrawn by December 1977. All locomotives were eventually broken up.

Very few external modifications existed within the class, except for some members being fitted with a revised cab ventilation system, which required a small covered grille on the cab-side corner posts.

The rebuilding of the ten redundant 'HA' Class 71 electric locomotives into dual-power Class 74 electro-diesel locomotives was performed by BR Crewe Works. The rebuilding was a major undertaking and was tantamount to building a new locomotive. These two views show the Erecting Shop with the left picture illustrating the new frames being assembled and the old cabs stored for re-use on the right. The illustration, right, shows the assembly line in an advanced stage with six locomotives taking shape.

Both: Author's Collection

The Class 74s were always used on the South Western Division's tracks radiating from Waterloo and their normal duties comprised of Waterloo – Southampton Docks and Waterloo – Bournemouth passenger/ van trains. On 8th July 1977 No. 74004 approaches Surbiton with the 07.30 Weymouth Quay – Waterloo boat train, which the electro-diesel would have operated from Bournemouth.

Colin J. Marsden

Front 3/4 study of the first locomotive of the fleet, No. E6101, rebuilt from electric locomotive No. E5015. The addition of waist height air and control jumpers permitted these locomotives to operate in multiple with each other, Class 73/1s, Class 33/1s and selected post-1951 EMUs. No. E6101 is seen in the works yard at Crewe.

Colin J. Marsden

One of the regular duties for the Class 74s was at the head of the frequent Waterloo – Southampton Docks Ocean Liner boat trains, the stock for which was normally stabled at Clapham Junction. In July 1976 No. 74002 is seen at Clapham Junction with empty stock bound for the adjacent yard, on its return to London after forming a down boat train.

Colin J. Marsden

There are very few illustrations available of Class 71 and 74 locomotives together, due to their different operating areas. However this view has been found showing electro-diesel No. E6108 and electric No. E5014 standing in the works yard at Eastleigh in 1968. The modification differences are quite noticeable.

Colin Boocock

Front 3/4 view of Class 74, taken from the No. 2 or diesel end. The nose-mounted air and control jumper connections are clearly visible in this view.

Colin J. Marsden

The main operating stronghold for the Class 74s was on the Waterloo – Southampton Docks boat trains, where the locomotives used their 650hp diesel engines to provide traction between Southampton and the dockside. No. 74010 poses in Southampton New Docks during September 1977.

Colin J. Marsden

In addition to working the main line services on the South Western section, the Class 74 locomotives were often used on empty stock diagrams between Waterloo and carriage sidings at Clapham Junction. On 18th September 1974 No. 74007 passes Vauxhall with the 09.57 Waterloo – Clapham Junction ecs train.

John Scrace

Class 76

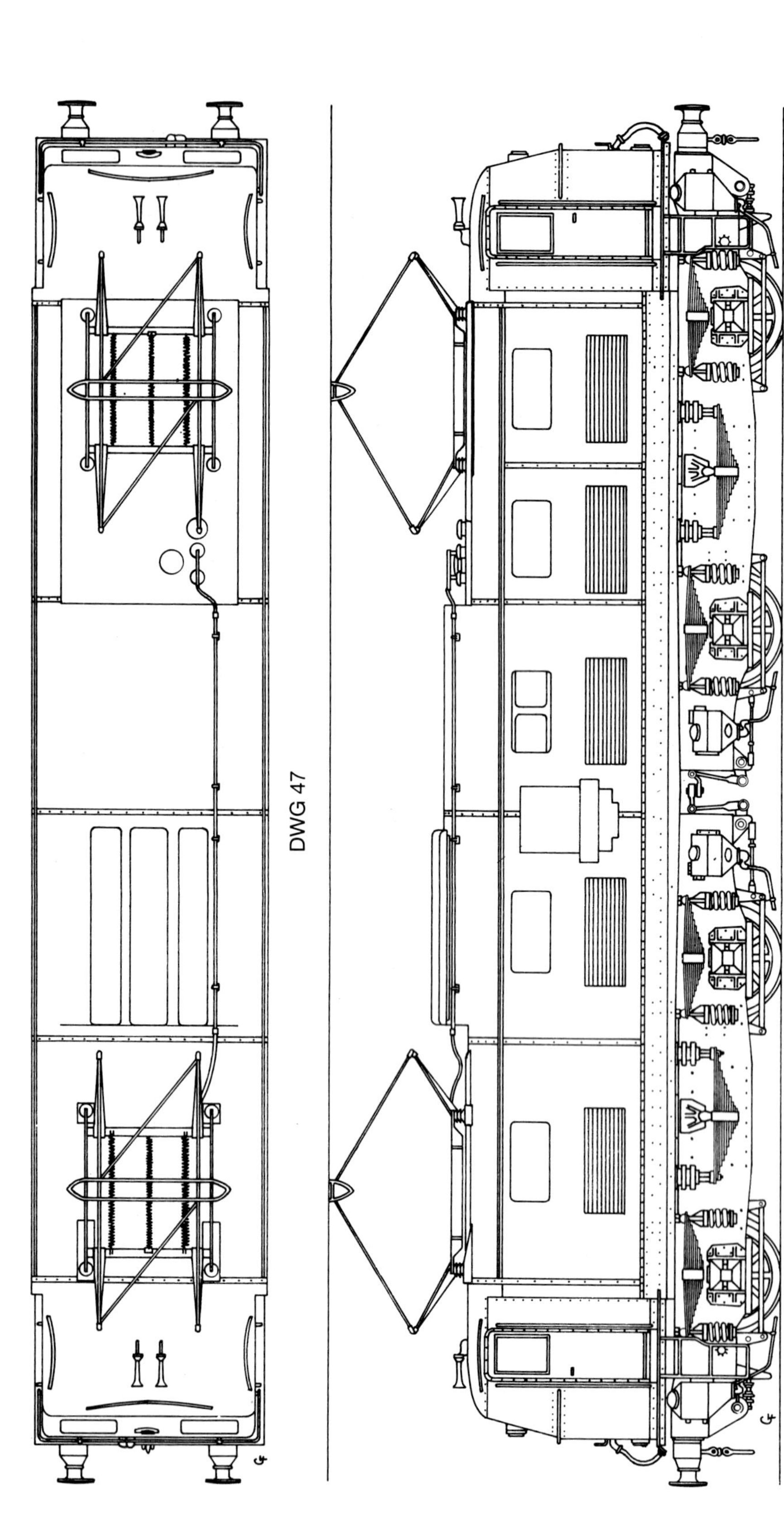

DWG 49

DWG 47
Roof detail of LNER prototype Bo-Bo locomotive No. 6000 (26000).

DWG 48
Side elevation of LNER prototype Bo-Bo locomotive No. 6000 (26000), showing pantographs in raised position.

DWG 49
Front end layout of LNER prototype locomotive No. 6000 (26000).

DWG 50
Class EM1/Class 76 roof detail.

DWG 51
Class 76 side elevation, showing side 'B'. This drawing applies to examples fitted with multiple control equipment.

DWG 52
Class 76 side elevation, showing side 'A'.

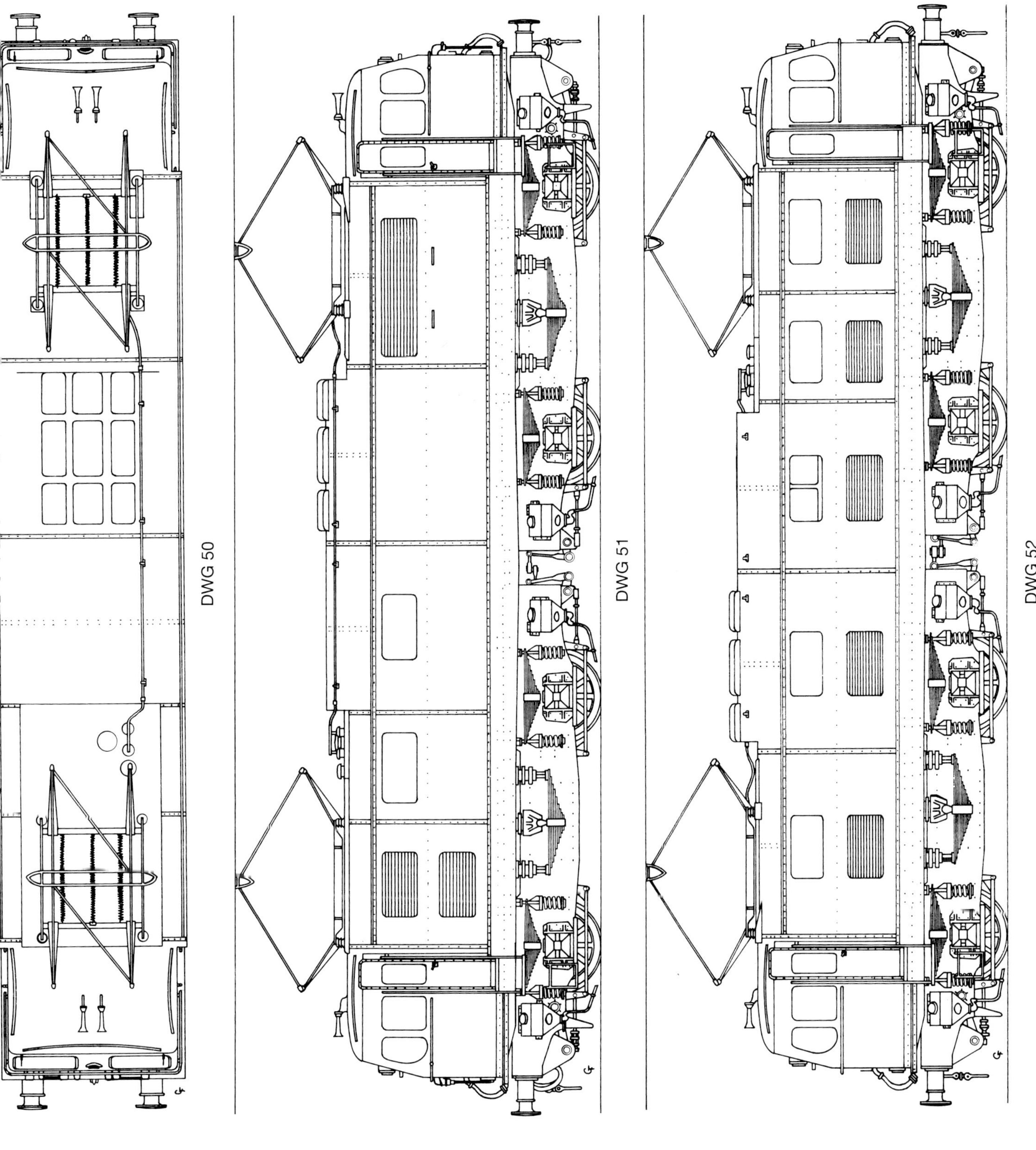
DWG 50
DWG 51
DWG 52

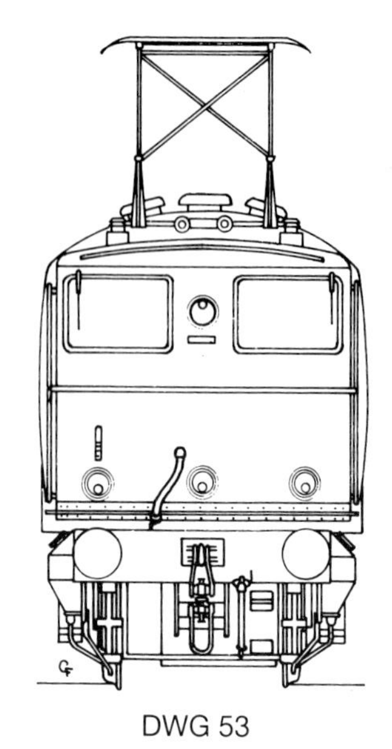

DWG 53

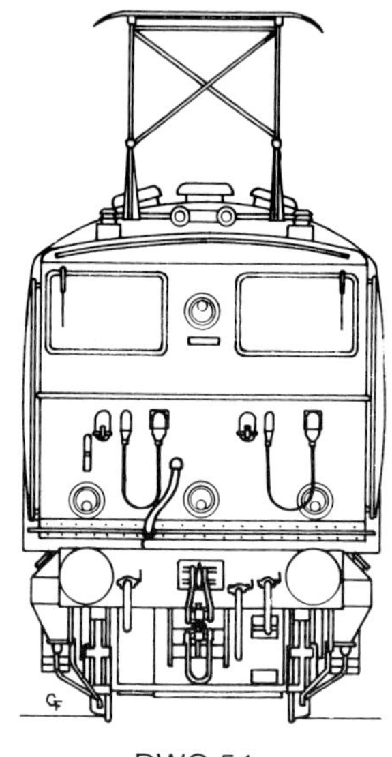

DWG 54

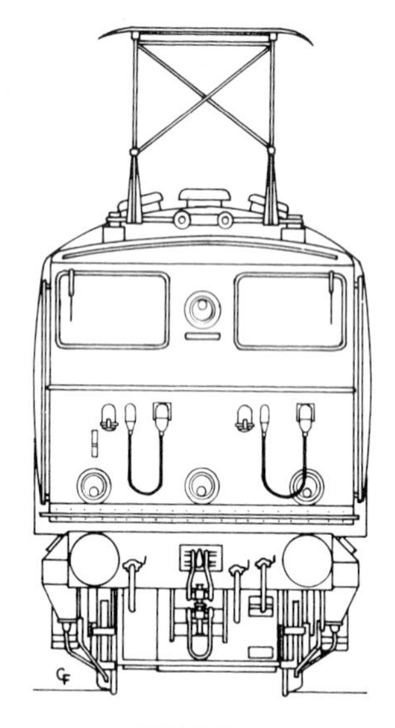

DWG 55

DWG 53
Class 76 front end layout, showing vacuum braking and steam heat pipe, as applicable to locomotives Nos E26046-57.

DWG 54
Class 76 front end layout, showing dual brake equipment and multiple control jumper cables/sockets.

DWG 55
Class 76 front end layout, showing air only braking system and multiple control jumper cables/sockets.

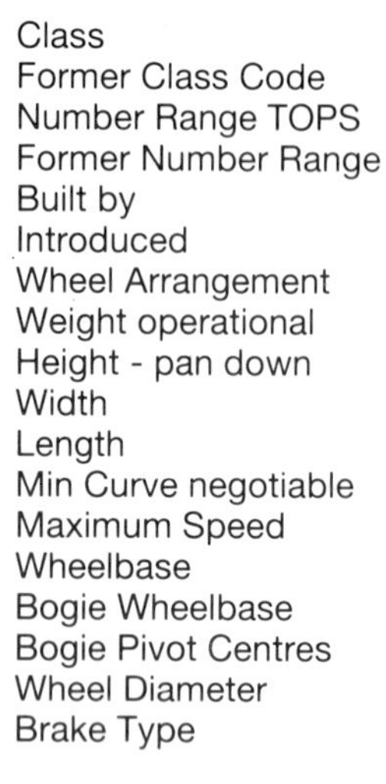

Class	76
Former Class Code	EM1
Number Range TOPS	76001-76057
Former Number Range	E26000-E26057 (Note: 1)
Built by	BR Doncaster and Gorton
Introduced	1941-53
Wheel Arrangement	Bo-Bo
Weight operational	88 tons
Height - pan down	13ft
Width	9ft
Length	50ft 4in
Min Curve negotiable	6 chains
Maximum Speed	65mph
Wheelbase	35ft
Bogie Wheelbase	11ft 6in
Bogie Pivot Centres	23ft 6in
Wheel Diameter	4ft 2in
Brake Type	Vacuum, Dual or air (Note: 2)
Sanding Equipment	Pneumatic
Heating Type	Steam - Bastian & Allen (Note: 3)
Route Availability	8
Coupling Restriction	Within class (Note: 4)

Brake Force	43 tons
Horsepower	1,868hp
Tractive Effort	45,000lb
Number of Traction Motors	4
Traction Motor Type	MV 186
Control System	Electro-Pneumatic
Gear Ratio	17:70
Pantograph Type	MV Cross Arm
Nominal Supply Voltage	1,500V dc overhead
Boiler Water Capacity	210gal (if fitted)

Note: 1 The original locomotive of this build was numbered 6701 when built in 1941. This was altered to 6000, and then into the main fleet.

Note: 2 This fleet was fitted with regenerative braking on the locomotive.

Note: 3 Steam heating equipment was only fitted to locomotives Nos 26000, 26046-26057 (76046-76057), and removed after passenger services stopped on the MSW line.

Note: 4 Multiple control facilities were fitted to locomotives Nos 26006-16/21-30 (76006-10/21-30).

It was the intention of the London & North Eastern Railway (LNER) for many years to invest in electrification of the cross-Pennine route between Sheffield and Manchester via Woodhead, as well as the branch from Penistone to Wath, and onwards from Sheffield to Rotherwood. Plans put forward during the 1920s/30s were gradually shelved and it was not until 1939 that financial approval was eventually given for the project, work starting almost immediately. The power system chosen for the scheme was 1,500V dc overhead. Concurrent with ground work commencing, the LNER works at Doncaster, under the auspices of Gresley, commenced production of a Bo-Bo electric locomotive. The locomotive bearing the LNER No. 6701 was completed in 1941 before the railway had any electrified tracks, therefore in order to test locomotive No. 6701 it was hauled across the Pennines to Manchester and used for a short period on the Manchester South Junction & Altrincham line, before returning to the LNER for storage.

After the end of world hostilities in 1945 the LNER were not able to recommend Pennine electrification immediately, and to avoid the prototype locomotive, by now renumbered 6000, laying idle any longer it was loaned to the Netherlands Railway in 1947.

By 1950, when electrification of the Pennine route was at an advanced stage, orders were placed for locomotives of two different types, classified EM1 and EM2. The EM1 fleet conformed almost identically to the original LNER locomotive and were constructed at the LNER works at Gorton near Manchester with electrical equipment supplied by Metropolitan-Vickers. A total of 57 EM1 locomotives were built, numbered 26001-26057. The fleet commenced operation in February 1952. At the end of the same year, the prototype locomotive was returned from Holland, and after extensive modification at Doncaster, entered service as EM1 No. 26000. This locomotive was named *Tommy*, a name bestowed upon it by European serviceman. After entry into service a total of twelve EM1 locomotives were named, all after creatures in Greek mythology.

From 1954 until July 1981, when the 1,500V dc route closed, this fleet operated the line, giving a good reliability figure. Over the years few modifications were carried out to the fleet, the most noticeable externally being the fitting of multiple control jumpers to 30 locomotives. Later in their careers several examples had their vacuum train brake equipment replaced by air brake only equipment.

Throughout its life the prototype locomotive remained identifiable by a slightly different front end and side design, which is documented in the accompanying drawings.

Under the BR five-figure TOPS renumbering system the EM1 fleet became Class 76 carrying the Nos 76001-76057. A small amount of renumbering occurred in later years when grouping of like brake fitted examples was made.

When built LNER No. 6701 was finished in LNER apple green livery which was carried until the early 1950s when black was applied. All production examples were completed in black which was later amended to BR locomotive green. After Autumn 1967 all repaints were carried out in standard BR blue with full yellow ends.

After closure of the 1,500V dc network all locomotives were sold for scrap, except No. 76020 (26020) which was saved by the National Railway Museum at York, who now have the locomotive painted in original lined black livery.

The pioneer LNER Bo-Bo 1,500V locomotive, built for the Manchester – Sheffield – Wath electrification poses in the works yard at Doncaster soon after construction in 1941. It will be noted that the locomotive is finished in LNER lined green livery and has one of its pantographs in a raised position.

Author's Collection

The 1950 order for Class EM1 locomotives was effected by Gorton Works, where the body shell of the first completed example is seen under high voltage electrical tests, which according to the sign on the front was carried out at 5,000V.
Author's Collection

When introduced the Class EM1 locomotives were painted in black livery, this later giving way to BR locomotive green, and subsequently, BR rail blue. Adorned in black livery and looking rather shabby, No. 26023 heads a heavy mineral train across the Pennines in 1957.

BR

During the 1960s traffic flows over the MSW electrified lines were intense with some 80 per cent of the locomotive fleet normally being rostered each day. On 14th July 1964 No. 26023 pauses in the loop at Penistone whilst sister locomotive, No. 26029, passes with a mineral train. Both locomotives are painted in green livery with a small yellow warning panel.
Colin J. Marsden

One of the main areas for 1,500V dc operation was around Penistone where the lines from Wath and Darnall parted company. Passing through the station on 24th July 1964 is No. 26054 at the head of the usual coal train. This locomotive was one of the batch equipped for passenger operation with steam heating, which is identified by the steam heat pipe on the buffer beam.
Colin J. Marsden

Many of the new 1,500V locomotives were, after completion, stored at Ilford depot in East London, with some locomotives operating trials in the London area using the ER(GE) 1,500V dc system. With only one pantograph in the raised position No. 26002, painted in black livery, stands at Shenfield on 12th November 1950 with the 12.18pm test train bound for Ilford.

L. Price

Throughout the 1970s and early 1980s a sizable row of Class 76 locomotives could be found each evening and at weekends stabled at Wath depot, a servicing shed used jointly by dc electric and diesel traction. This general view of the shed taken in 1977 shows nine Class 76 locomotives stabled in one line, while two Class 37s and a Class 47 stand outside the shed.

Colin J. Marsden

It was not uncommon to find the Class 76 locomotives operating in pairs over the MSW route at the head of long and heavy freight trains. On 3rd September 1980, Nos 76025 and 76016 pull away from the photographer at Valehouse with a westbound coal train.

Brian Morrison

The locomotives fitted with multiple control jumpers were those usually used in pairs, as the two locomotives could be controlled by one driver. A pair of mu fitted examples, equipped with air-only train brake equipment, approach Valehouse with a westbound merry-go-round train. The leading locomotive is No. 76010.

Brian Morrison

Class 77

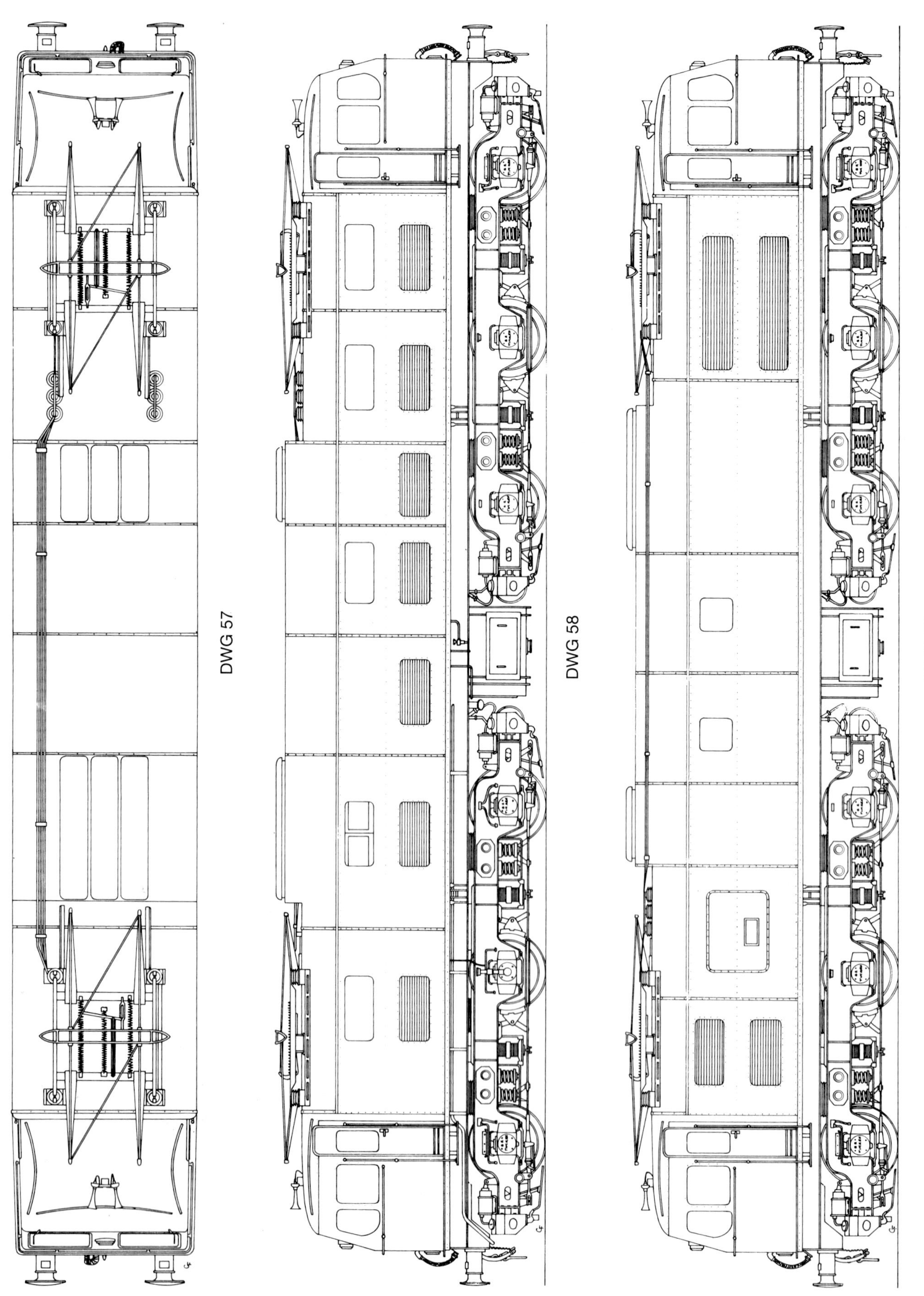

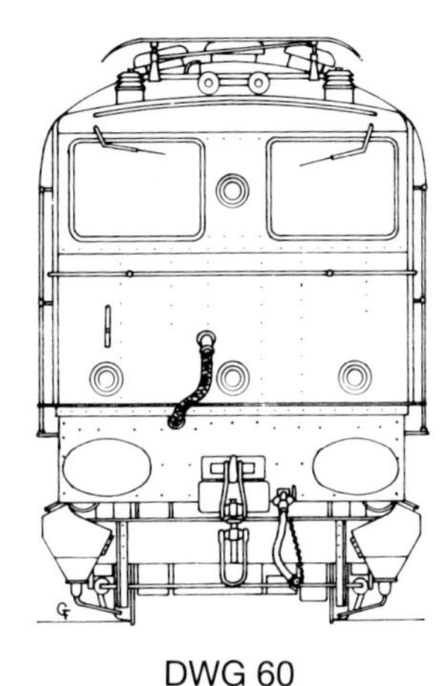

DWG 60

DWG 57
EM1/Class 77 roof detail.

DWG 58
EM1/Class 77 side 'A' elevation. No. 1 end on left.

DWG 59
EM1/Class 77 side 'B' elevation. No. 1 end on right.

DWG 60
EM1/Class 77 front end elevation.

Class	77
Former Class Type	EM2
Number Range	E27000-E27006
Built by	BR Gorton
Introduced	1953-54
Wheel Arrangement	Co-Co
Weight	102 tons
Height - pan down	13ft
Width	8ft 10in
Length	59ft
Min Curve negotiable	6 chains
Maximum Speed	90 mph
Wheelbase	46ft 2in
Bogie Wheelbase	15ft 10in
Bogie Pivot Centres	30ft 6in
Wheel Diameter	3ft 7in

Brake Type	Vacuum (Note: 1)
Sanding Equipment	Pneumatic
Heating Type	Steam - Bastian & Allen
Route Availability	8
Coupling Restriction	Not multiple fitted
Horsepower	2,300hp
Tractive Effort (maximum)	45,000lb
Number of Traction Motors	6
Traction Motor Type	MV 146
Control System	Electro-Pneumatic
Gear Ratio	17:64
Pantograph Type	MV Cross Arm
Nominal Supply Voltage	1,500V dc Overhead

Note: 1 This fleet were fitted with regenerative braking on the locomotive.

Following early experience with the prototype LNER 1,500V dc locomotive, when the ride characteristics gave cause for concern, it was decided to opt for a Co-Co locomotive for use on passenger duties for the Trans-Pennine electrified route. Under original plans it was envisaged that a fleet of 27 Co-Co locomotives would be required, however in light of operational experience with the EM1 Bo-Bo fleet and subsequent bogie modifications, which permitted a speed increase, the production fleet of Co-Co locomotives was reduced to just seven. Under the electric locomotive classification system this fleet became Class EM2, which was later amended to BR Class 77. The number range allocated was E27000-E27006.

Power and control equipments were supplied by Metropolitan Vickers, while construction was effected by the Gorton Works of BR in 1953-54. One of the major structural changes on this fleet to the earlier EM1 type was the fitting of the buffing and draw gear onto the bodywork and not the bogie frame.

From their introduction this small fleet of main line electric locomotives were always deployed on the Manchester – Sheffield – Wath line. However, when during the late 1960s, BR decided to withdraw the passenger services, and re-route the line's passenger duties via the non-electrified Hope Valley route, the class was withdrawn. Although removed from service there was plenty of life left in these machines and BR offered the entire fleet for sale. After only a short period on the market the Netherlands Railway (NS) became interested and purchased all locomotives, which were shipped to Holland and rebuilt to their requirements. To cover NS operations only six locomotives were needed, the seventh, No. E27005 being broken up for spares. Once in service the six NS locomotives were allocated numbers in the 1501 – 1506 series and proved very reliable, remaining in traffic until late 1986. Two locomotives of the fleet have been saved by the preservation movement and returned to England. No. E27000 *Electra* is currently stored at Ilford, while No. E27001 is on display at the Greater Manchester Museum of Science & Industry.

When in BR service the locomotives were painted in Electric blue when constructed, this later being amended to BR green. All seven examples of the class were named in association with gods of mythology.

The larger and more powerful Class EM2 locomotives were designed for the Trans-Pennine passenger services and were indeed seldom seen on freight duties. No. 27004 *Juno* is seen near Penistone in June 1964 at the head of a Manchester – Sheffield Victoria service.

Author's Collection

The fleet of seven Class EM2, later BR Class 77 locomotives, were constructed at the former LNER works at Gorton, near Manchester, and received all their classified maintenance at the same location. Painted in BR green livery No. 27001 *Ariadne* is seen outside Gorton Works in 1966.

Colin J. Marsden

Displaying the main line black livery No. 27003, later to be named *Diana,* slowly pulls the 2.10pm service to Marylebone out of Manchester London Road on 18th March 1955.

John Faulkner

With a splendid array of BR and pre-Nationalisation stock behind, No. 27001, still un-named, heads a Manchester – Sheffield working near Oughty Bridge in the summer of 1957.

Author's Collection

During the early 1960s, in the railway's attempt to improve its aesthetic image, new liveries were applied, including a shade known as electric blue, to many electric locomotives. No. 27002 *Aurora* is seen outside Reddish depot after repainting into the new scheme.

L. Price

One of the major exhibits at the British Railway's Traction Exhibition at Willesden in May/June 1954 was brand new No. 27002, delivered direct from the builder's works at Gorton. With pantograph raised the locomotive is seen on 27th May 1954.

BR

As detailed in the introductory text, after their useful life in Britain and their premature withdrawal, the entire fleet was sold to the NS – Nederlandse Spoorwegen – where the locomotives were rebuilt to NS standards at Utrecht. The former No. 27000 *Electra*, disguised as NS No. 1502, is seen at Hoorn on 15th March 1986 with the RT&P 'EM2 Electra Special' from Hook of Holland, Utrecht and Amsterdam.

John Tuffs

After their demise on NS two examples were returned to England for preservation, NS Nos 1502 and 1505, the former BR Nos 27000 *Electra* and 27001 *Ariadne*. NS No. 1502 is seen on display at the Basingstoke Rail Event on 26/27th September 1987.

Colin J. Marsden

In the autumn of 1988 No. 27000 *Electra* was repainted by its owners, the EM2 Preservation Society, into BR green livery. The locomotive is seen at London Waterloo on display at the Ian Allan Network Day on 1st October 1988.

Colin J. Marsden

Class 80

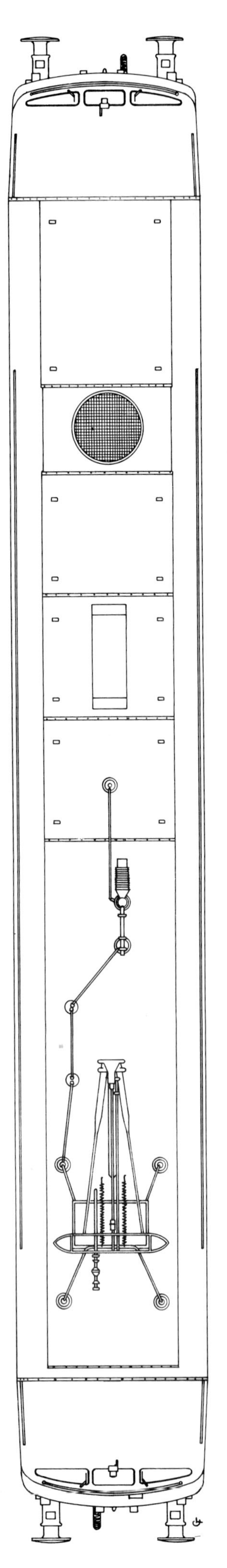
DWG 61

DWG 61
BR/MV prototype 25kV electric locomotive No. E1000 (E2001) roof detail.

DWG 62
BR/MV prototype 25kV electric locomotive No. E1000 (E2001) side elevation.

DWG 63
BR/MV prototype 25kV electric locomotive No. E1000 (E2001) front end detail.

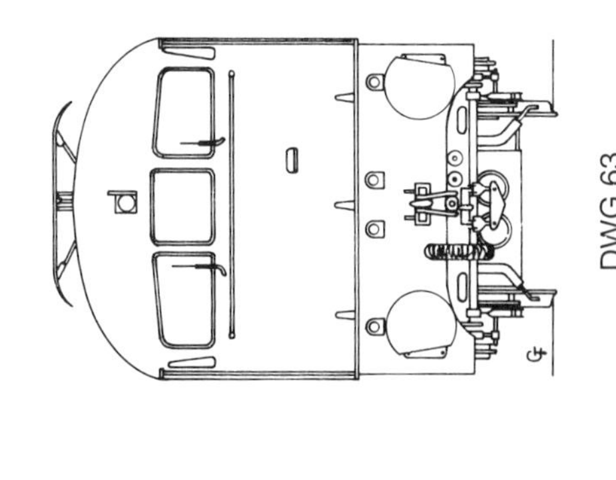
DWG 63

DWG 62

Class	80
Number	E2001
Original 1957 Number	E1000
Former Number	18100 (Note: 1)
Rebuilt by	Metropolitan – Vickers
Introduced (Original)	1952
Introduced Rebuilt	1958
Wheel Arrangement	A1A-A1A
Weight	109 tons
Height - pan down	12ft 10in
Width	8ft 8¼in
Length	66ft 9¼in
Min Curve negotiable	4 chains
Maximum Speed	90mph
Wheelbase	53ft
Bogie Wheelbase	15ft
Bogie Pivot Centres	37ft 6in
Wheel Diameter	3ft 8in
Brake Type	Vacuum
Sanding Equipment	Pneumatic
Heating Type	Electric
Route Availability	7
Multiple Coupling Restriction	Not multiple fitted
Horsepower	2,500hp
Tractive Effort	40,000lb
Number of Traction Motors	4
Traction Motor Type	MV
Control System	LT Tap Changing
Gear Drive	Direct spur, single reduction
Gear Ratio	21:58
Pantograph Type	Stone-Faiveley
Rectifier Type	Mercury Arc
Nominal Supply Voltage	25kV ac

Note: 1 This locomotive was rebuilt from prototype gas-turbine No. 18100, and used for training on the LM 25kV AC system.

The first of the 25kV ac overhead electric classes was in many ways an unusual locomotive, being originally built as a gas-turbine by Metropolitan-Vickers in 1952. As a gas-turbine, No. 18100 the locomotive was operated on the WR until being made redundant in 1958.

After electrification of the London Midland Region main line was authorised and traction orders placed, it became apparent that there would be a long wait before any operational hardware would be available. Following much deliberation it was decided to contract Metropolitan – Vickers to rebuild the former gas-turbine No. 18100 into a 25kV ac electric locomotive, enabling training and overhead equipment testing to take place at an early date. At the time of the conversion decision, No. 18100 lay dumped at Dukinfield near Manchester, from where it was hauled to the MV works at Stockton-on-Tees. The rebuilding work was major and consisted of the removal of the former gas-turbine power unit, auxiliary combustion equipment, dc power equipment, fuel tanks and control equipment. In their place ac power, control, transformer and rectifier units were installed. The cabs were also heavily rebuilt to remove the previously fitted GWR style right-hand driving layout. The roof also had to be modified to accommodate the pantograph. Another structural modification worthy of note was the trimming of the buffers to bring the machine within the required gauge in terms of width. To provide traction power for the new electric locomotive, four of the original six traction motors were retained, as were some of the auxiliary machines such as traction motor blowers, vacuum brake exhauster, air compressor and cooling equipment. During the rebuilding work a very small staff room was incorporated at No. 1 end, which was intended as a training classroom.

The pioneer LM ac electric locomotive, finished in main line black livery off-set by a silver body band, was released from Metropolitan-Vickers in Autumn 1958, still carrying its gas-turbine number 18100. After initial testing in the Styal area the machine was renumbered to E1000 and put to work on the Manchester – Crewe line between Mauldeth Road and Wilmslow. After a short period as No. E1000 the locomotive was renumbered to E2001.

After some twelve months of being the only 25kV ac electric locomotive in service, the production classes started to enter traffic spelling the end for No. E2001, and after mid-1961 the locomotive saw little use. During the autumn of 1961 No. E2001 was sent north to Scotland where it was used on the Glasgow electrified area for equipment testing, but by Christmas the locomotive was returned to the London Midland Region, being stored at various locations such as Crewe, Goostrey and finally Rugby. For a period in 1964 No. E2001 did see some further use, when deployed as a training locomotive at Rugby, but after only a few months the machine was again stood down. By early 1968 the locomotive was deleted from stock, but lay for an extended period in sidings at Market Harborough and Rugby before being sold to J. Cashmore Ltd of Tipton for scrap in 1972.

Few physical body alterations were carried out to gas-turbine locomotive No. 18100 when its was rebuilt into 25kV prototype locomotive No E1000, later E2001. This view shows the locomotive after conversion, with the pantograph at the far end. Note the empty underframe between the bogies, and the trimmed buffers.

GEC Traction

No. E1000 (E2001) only operated as an electric locomotive for a short period, as when sufficient numbers of production 25kV locomotives were available all training and testing was concentrated on these types. No. E1000 is seen heading a rake of Mk 1 stock on the Styal line during 1961.

Author's Collection

An illustration taken only a few days after release from the Metropolitan-Vickers works, shows the prototype electric locomotive lined up for an official photograph, although still carrying its original gas-turbine number, 18100.

Author's Collection

Former gas-turbine locomotive No. 18100 cab layout, as revised for electric operation. The train and locomotive brake valves are located on the left, with the power controller in the middle foreground.

Colin J. Marsden

Class 81

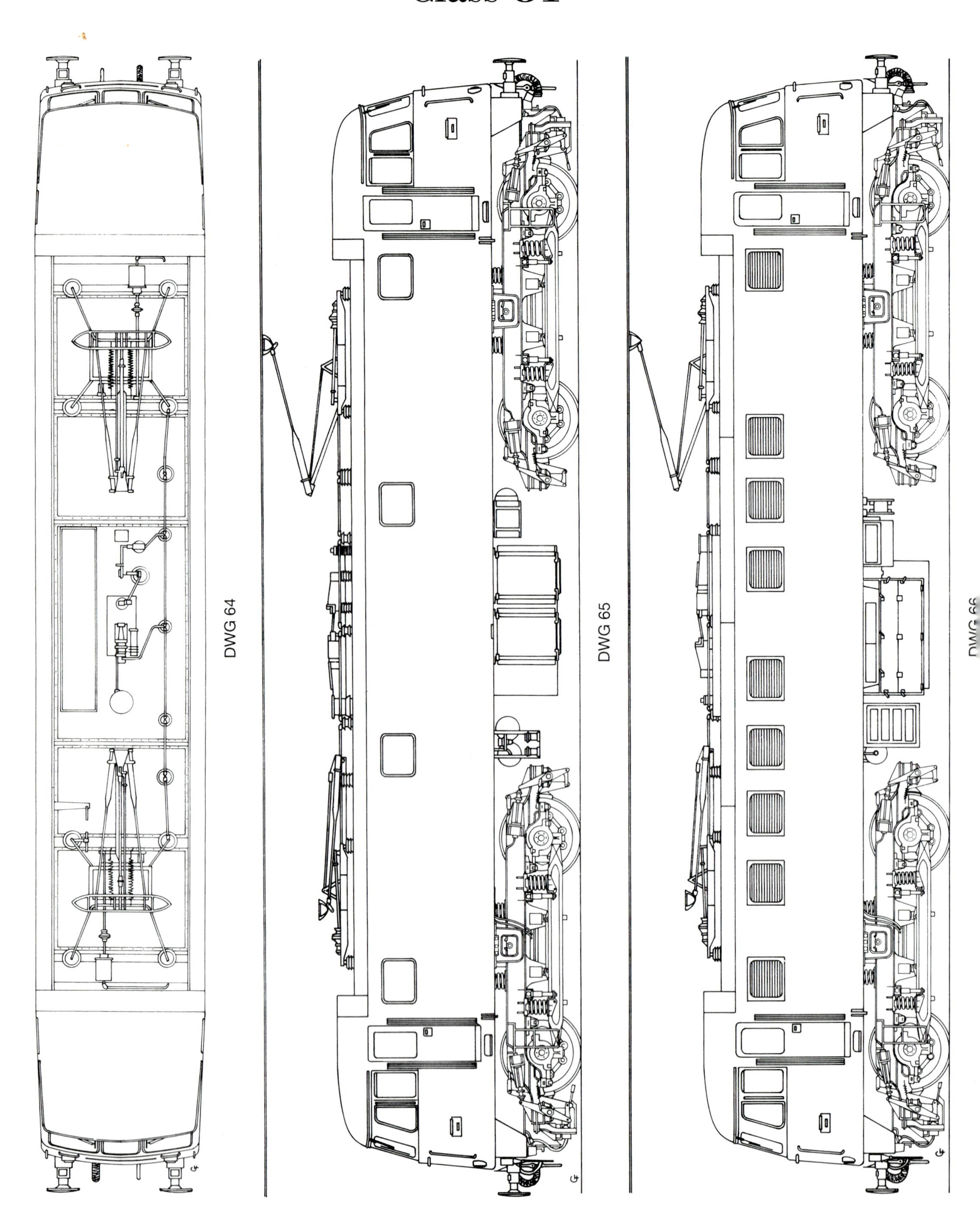

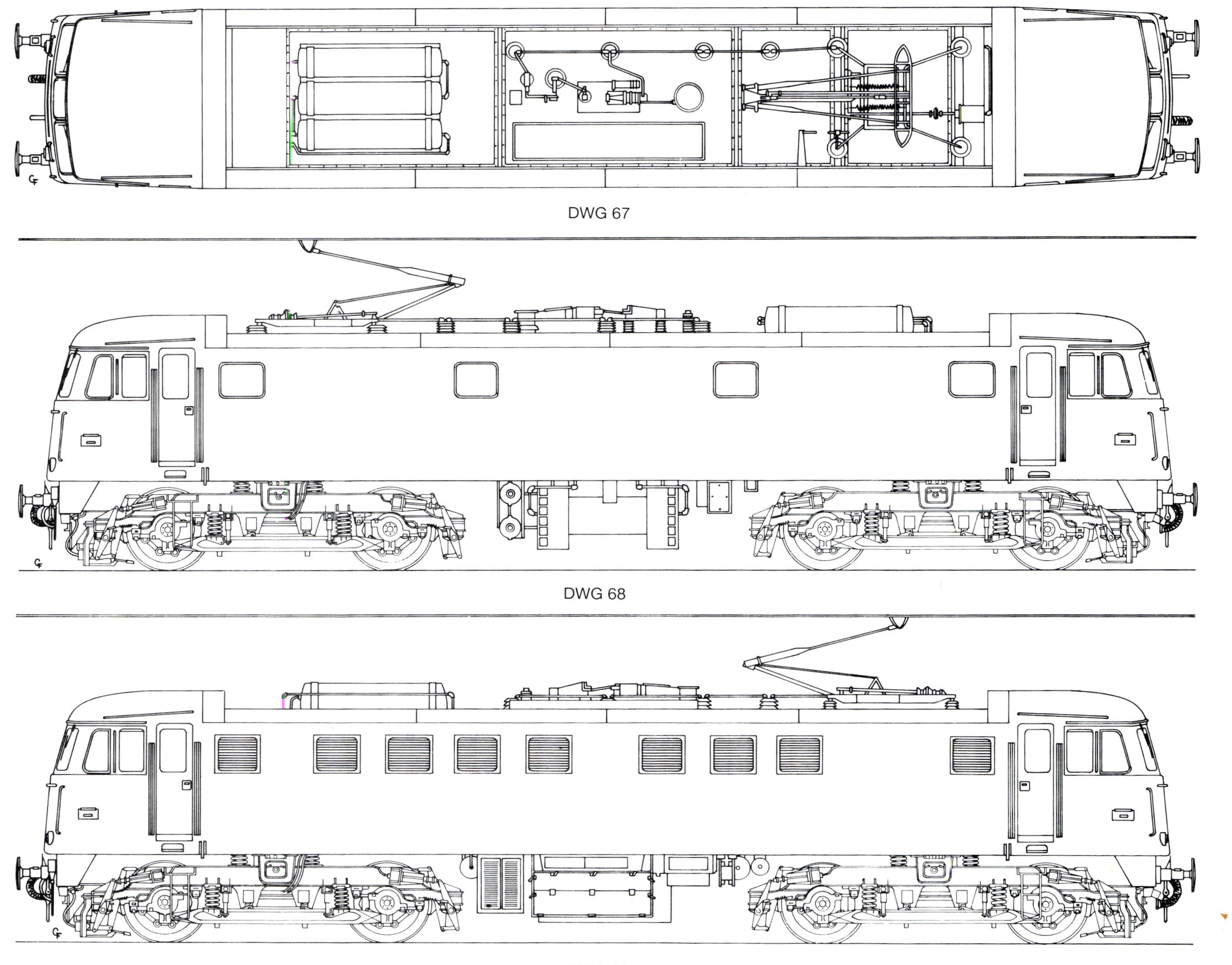

DWG 67

DWG 68

DWG 69

DWG 70

DWG 71

Class	81
Former Class Code	AL1
Number Range (TOPS)	81001-81022
Former Number Range	E3001-E3023, E3096-E3097 (E3301-E3302)
Built by	BRC&W Ltd
Introduced	1959-64
Wheel Arrangement	Bo-Bo
Weight	79 tons
Height - pan down	13ft $0^{9}/_{16}$in
Width	8ft $8^{1}/_{2}$in
Length	56ft 6in
Min Curve negotiable	4 chains
Maximum Speed	100mph (Note: 1)
Wheelbase	42ft 3in
Bogie Wheelbase	10ft 9in
Bogie Pivot Centres	31ft 6in
Wheel Diameter	4ft
Brake Type	Dual (Note: 2)
Sanding Equipment	Pneumatic
Heating Type	Electric - Index 66
Route Availability	6
Coupling Restriction	Not multiple fitted
Brake Force	40 tons
Horsepower (continuous)	3,200hp
(maximum)	4,800hp
Tractive Effort (maximum)	50,000lb
Number of Traction Motors	4
Traction Motor Type	AEI 189
Control System	LT Tap Changing
Gear Drive	Alsthom Quill, single reduction
Gear Ratio	29:76
Pantograph Type	Stone-Faiveley
Rectifier Type	Silicon (Note: 3)
Nominal Supply Voltage	25kV ac

Note: 1 The maximum speed of this class was reduced to 80mph in 1986.

Note: 2 When built vacuum only brakes were fitted.

Note: 3 When built mercury arc rectifiers were fitted.

DWG 64
AL1/Class 81 roof detail, showing as-built condition with two pantographs.

DWG 65
AL1/Class 81 side 'B' elevation, showing as-built condition with two pantographs. No. 1 end on right.

DWG 66
AL1/Class 81 side 'A' elevation, showing as-built condition with two pantographs. No.1 end on left.

DWG 67
Class 81 roof detail, after refurbishment with one pantograph. No. 1 end on right.

DWG 68
Class 81 side 'A' elevation, after refurbishment with only one pantograph, and angled rain water strip on cab side roof. No. 1 end on right.

DWG 69
Class 81 side 'B' elevation, after refurbishment with only one pantograph, and angled rain water strip on cab side roof. No. 1 end on left.

DWG 70
AL1/Class 81 front end layout, showing two pantographs and vacuum only train brake equipment.

DWG 71
Class 81 front end layout, showing revised layout with dual brake equipment and modified route indicator panel.

The pioneer order for 25kV ac 'production' locomotives was spread amongst several major builders. Numerically the first series No. E3001 onwards, was the AL1 fleet which consisted of 23 Type 'A' (passenger), and two Type 'B' (freight) locomotives. Construction was carried out by the Birmingham RC&W Co., who were acting as chief sub-contractor to AEI. The mechanical portion, designed by BRC&W, was a load bearing structure, being formed of girder steel, plated in medium gauge sheet. The body structure had full width cabs at either end, and housed all electrical equipment in lockable compartments in the between cab section on one side of the body, the other side housing a between cab walkway. The between cab roof section was lower than standard height to accommodate the pantographs, which were electrically arranged to be able to receive power at either 6.25kV or 25kV depending on the area in which the locomotive was used, although the 6.25kV system was never used.

The first of the production ac classes, the AL1s, started to emerge at the end of 1959 when No. E3001 was released from the BRC&W works at Smethwick, and transferred to the Styal line for active testing. This view shows No. E3001 a few days after release from BRC&W prior to testing. Note that only one pantograph, that for 25kV operation, is raised, and that one of the ventilation louvres at the far end is of a modified type.

Author's Collection

The two body sides of the Class AL1, later Class 81, were totally different, one having four windows, and the other accommodating nine ventilation grilles. When introduced the Type 'A' locomotives were numbered in the E3001-E3023 range, while the Type 'B' locomotives were numbered E3301-E3302, later amended to E3096-E3097. Under the TOPS renumbering system the entire fleet became Nos 81001-81022.

When constructed the livery applied was Electric blue, with white cab roof and window surrounds. Over the years yellow warning panels were progressively added, and later BR standard Rail blue was carried with full yellow ends.

When introduced the class were fitted for vacuum train brake operation only, this being supplemented by air brake equipment in the late 1960s, at which time the redundant 6.25kV pantograph was removed.

Until the introduction of second generation electric classes during the mid-1960s the AL1s, in company with the other pilot electric types, were to be found operating over all parts of the LM electrified network, heading both passenger and freight traffic. In latter years members were allocated to Glasgow Shields Road depot, however their diverse operations took them all over the electrified network right to the end of there operating careers. Major withdrawal of the fleet commenced during 1988, with the fleet being eliminated by mid-1991.

When built the Class AL1s had four-character route indicator panels. These were replaced by black screens and white marker 'cut-outs' during the mid-1970s. In later years the redundant indicator boxes were plated over and two sealed beam marker lights fitted.

Although the Class 81s are now a part of railway history, one member of the class, No. E3003 (81002), has been preserved at the Railway Age, Crewe.

From their introduction the ac classes have received the majority of their classified overhauls at the BR/BREL workshops at Crewe where special facilities for their maintenance was provided. This mid-1960's view shows four Class 81s and a Class 40 diesel receiving maintenance at Crewe.

BR

In common with all main line locomotives yellow high-visibility warning panels and later, full yellow ends were progressively applied from the mid-1960s. Whilst adorned with just a small yellow warning panel most locomotives retained a white cab roof which gave a very pleasing appearance. Hauling a rake of Mk1 and Mk2 stock No. E3003 is seen heading south on the West Coast Main Line in October 1968.

BR

Traversing the up slow line at Wolverton, Class 81 No. 81017 heads a Bescot – Willesden Yard 'Speedlink' freight service on 27th July 1988. In the background part of Wolverton Works can be seen.

Michael J. Collins

Displaying the standard BR rail blue livery No. 81007 is seen at Euston with a rake of empty coaching stock on 3rd October 1981. By this time the locomotive had been refurbished with only one pantograph (at the far end), and dual brake equipment.

Colin J. Marsden

The Class 81 fleet finished their working days allocated to Glasgow Shields depot, being used on all sections of the LM/Sc electrified network. On 28th April 1984 No. 81017 fitted with sealed beam marker lights, is seen at Stafford with the 13.26 relief from Euston to Glasgow Central.

Michael J. Collins

Class 82

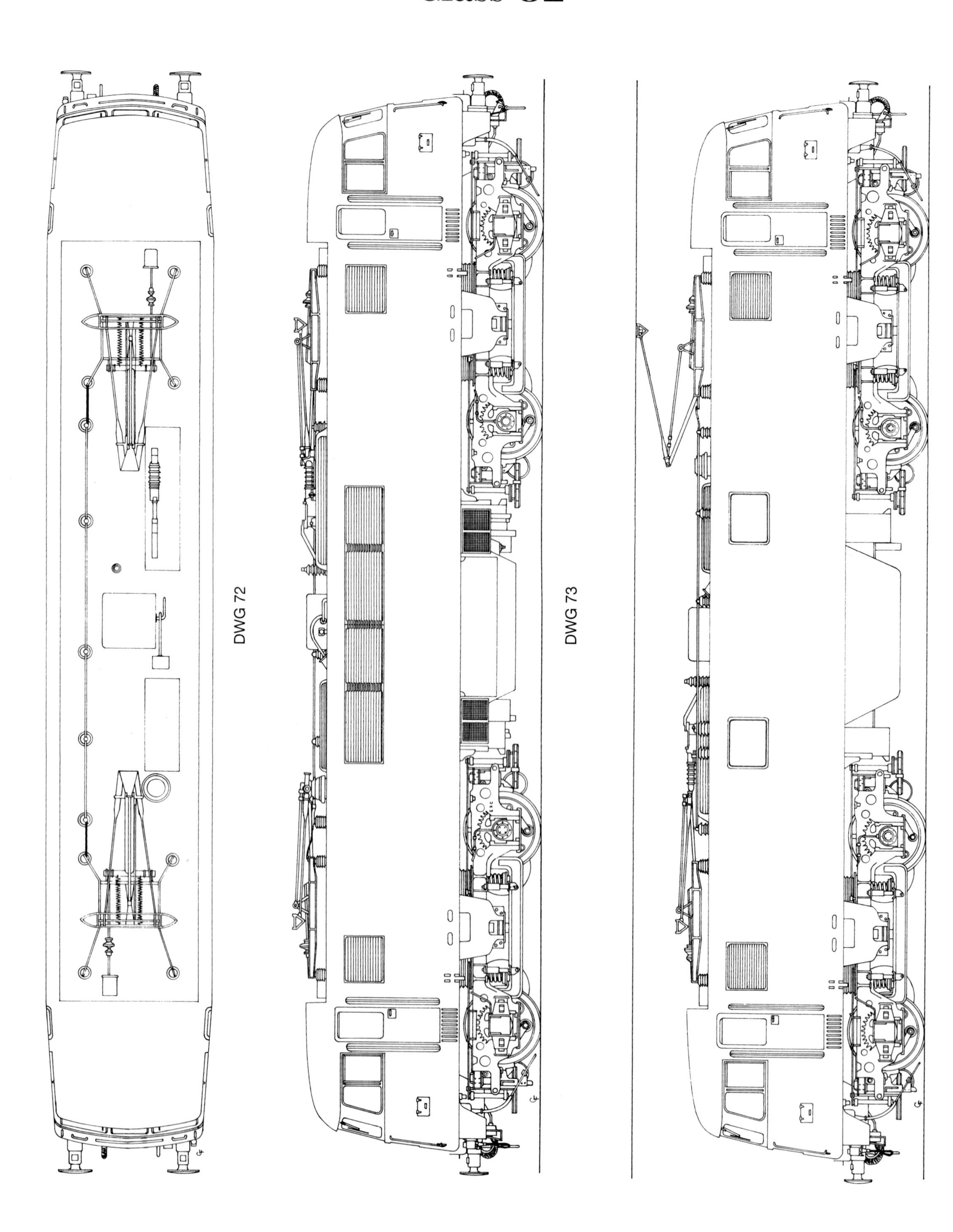

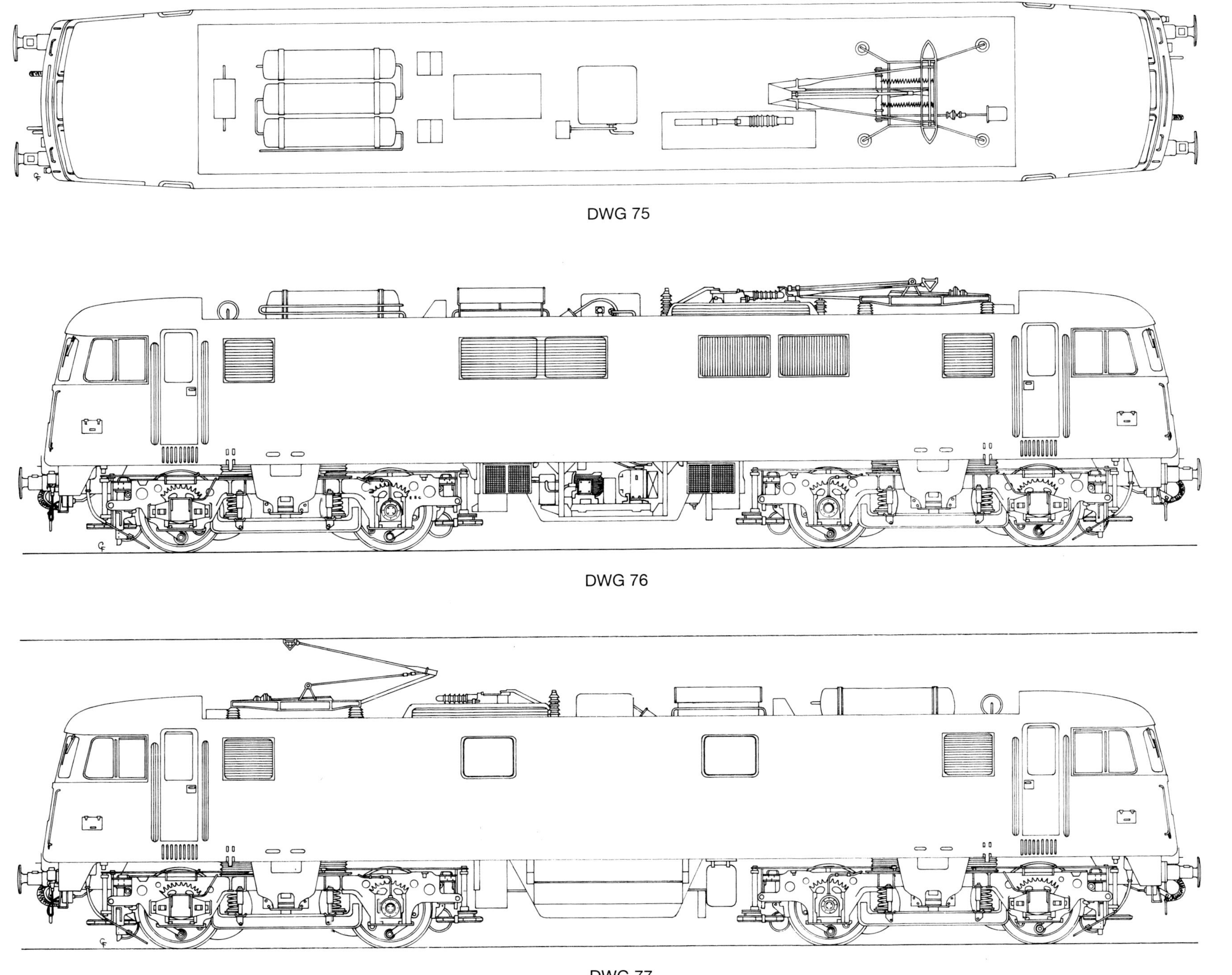

DWG 75

DWG 76

DWG 77

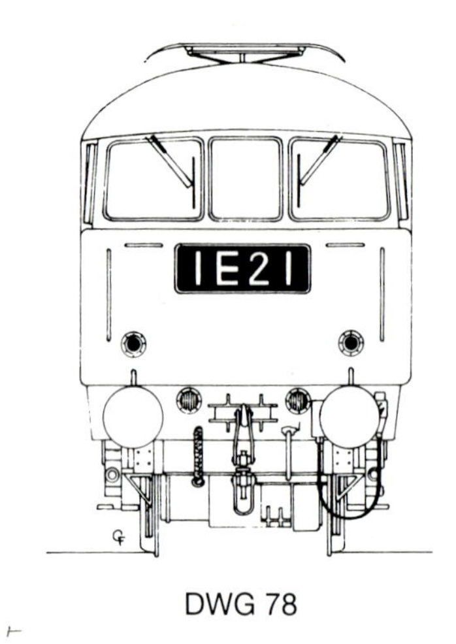

DWG 78

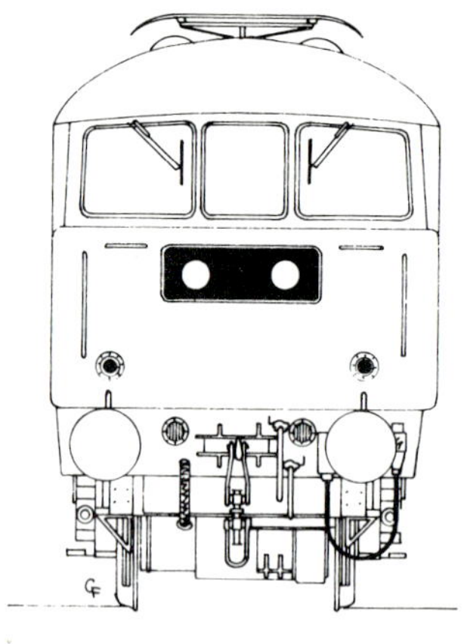

DWG 79

Class	82
Former Class Code	AL2
Number Range (TOPS)	82001-82008
Former Number Range	E3046-E3055
Built by	Beyer, Peacock
Introduced	1960-61
Wheel Arrangement	Bo-Bo
Weight	80 tons
Height - pan down	13ft $0\frac{5}{8}$in
Width	8ft 9in
Length	56ft
Min Curve negotiable	4 chains
Maximum Speed	100mph (Note: 1)
Wheelbase	40ft 9in
Bogie Wheelbase	10ft
Bogie Pivot Centres	30ft 9in
Wheel Diameter	4ft
Brake Type	Dual (Note: 2)
Sanding Equipment	Pneumatic
Heating Type	Electric - Index 66
Route Availability	6
Coupling Restriction	Not multiple fitted

Brake Force		38 tons
Horsepower	(continuous)	3,300hp
	(maximum)	5,500hp
Tractive Effort	(maximum)	50,000lb
Number of Traction Motors		4
Traction Motor Type		AEI 189
Control System		HT Tap Changing
Gear Drive		Alsthom Quill, single reduction
Gear Ratio		29:76
Pantograph Type		Stone-Faiveley
Rectifier Type		Silicon (Note: 3)
Nominal Supply Voltage		25kV ac

Note: 1 Class 82 locomotives remaining in service after 1986 were restricted to 40mph for ecs duties in the London area.

Note: 2 When introduced the Class 82s were fitted for vacuum braking only.

Note: 3 When built mercury arc rectifiers were fitted.

DWG 72
AL2/Class 82 roof detail, showing as-built condition with two pantographs.

DWG 73
AL2/Class 82 side 'A' elevation, showing as-built condition with two pantographs, No. 1 end on left.

DWG 74
AL2/Class 82 side 'B' elevation, showing as-built condition with two pantographs, No. 1 end on right.

DWG 75
Class 82 roof detail, after refurbishment with one pantograph. No. 1 end on left.

DWG 76
Class 82 side 'A' elevation, after refurbishment with revised louvre position. No. 1 end on left.

DWG 77
Class 82 side 'B' elevation, after refurbishment with one pantograph. No. 1 end on right.

DWG 78
AL2/Class 82 front end detail, showing the as-built condition with vacuum only train braking.

DWG 79
Class 82 front end detail, showing dual brake fitment.

Under the original ac orders a fleet of ten Type 'A' (passenger) locomotives was ordered from AEI/Metropolitan-Vickers, who sub-contracted mechanical construction to Beyer, Peacock of Gorton, Manchester. Under the ac locomotive classification system this fleet became Class AL2 and allocated numbers in the range E3046-E3055. Under the later BR TOPS numbering system the fleet became Class 82, being numbered 82001-82008. The first locomotive of the build emerged in May 1960, and immediately took up trial running on the Styal line in Manchester.

The external appearance of this fleet closely followed the style of the previously detailed Class AL1 design, but the construction method was significantly different, incorporating a separate underframe and body. At the design stage it was envisaged that weight might be something of a problem, and to overcome this much alloy and glass fibre was used.

The between cab layout was almost identical to the Class AL1, and the cab layout conformed to the BTC standard style adopted for all the ac builds of the 1960s. Two pantographs were again fitted from new, as was vacuum only train brake equipment. During the early 1970s, when air braking was being more frequently introduced, the entire fleet were refurbished to provide dual brake equipment, at the same time the redundant 6.25kV pantograph was removed and the space taken by the additional air braking reservoirs.

When constructed, one side of the locomotive had six louvred covered openings, while the other had two windows and two louvred panels. After refurbishing during the early 1970s the louvred side was largely altered to incorporate an additional vent panel to improve internal ventilation.

After introduction the AL2 fleet was allocated to Longsight depot in Manchester, and displayed the standard Electric blue livery, offset by white cab roofs and cab window surrounds. Over the years standard Rail blue livery with full yellow warning ends was applied.

The AL2 fleet performed extremely well on the West Coast Main Line and were well liked by the train crews. However by 1982-83 the majority of the fleet were deemed as surplus to requirements and stored, eventually being withdrawn. However Nos 82005 and 82008 were retained until 1987 for empty stock duties in the London area, being allocated to Willesden.

Once sufficient electric traction was available electric services were introduced progressively as the various routes were energised. Taken only a short time after the switching on of the Crewe – Liverpool section on 1st January 1962 an electrically hauled Inter-regional express approaches Liverpool Lime Street headed by a Class AL2 and an AL1 locomotive. *BR*

6th March 1967 was a big day in the annals of BR modernisation. During that morning the Official Opening of Birmingham New Street station took place, coupled with the introduction of the new high speed electric timings between London, Birmingham and the Midlands, bringing to an end the eight-year modernisation programme. One of the first workings to the new schedule was the 10.27 Manchester Piccadilly – Euston via Birmingham, seen here leaving New Street behind Class 82 No. E3054.

Author's Collection

The Class 82 locomotives in their refurbished state provided the LM with a reasonably satisfactory fleet of power units, and indeed prior to the full availability of Class 86 and 87 locomotives, were to be found at the head of crack InterCity services. On 25th October 1981 No. 82005 passes Castlethorpe with the Sundays Only 10.50 Euston – Liverpool service.

Michael J. Collins

With its No. 2 or pantograph end leading, and the locomotive's 'B' side nearest the camera, No. 82008 arrives at Willesden Yard on 12th May 1980 with empty newspaper vans from Birmingham.
Colin J. Marsden

During the early years of ac passenger operation on the WCML, double heading of services was not uncommon. Here, Class AL2 No. E3047 pilots Class AL5 No. E3073 at Stafford with an express service bound for London.
Norman E. Preedy

Class 82 No. 82002 used for ecs duties in the London area was repainted into InterCity livery to match the coaching stock during the late 1980s. Looking rather dilapidated after passing through the acid coach washer at Willesden Brent Sidings the locomotive is seen posed inside Willesden DED. Note the cab-shore telephone aerial on the front end to the right of the driver's window.
Colin J. Marsden

Class 83

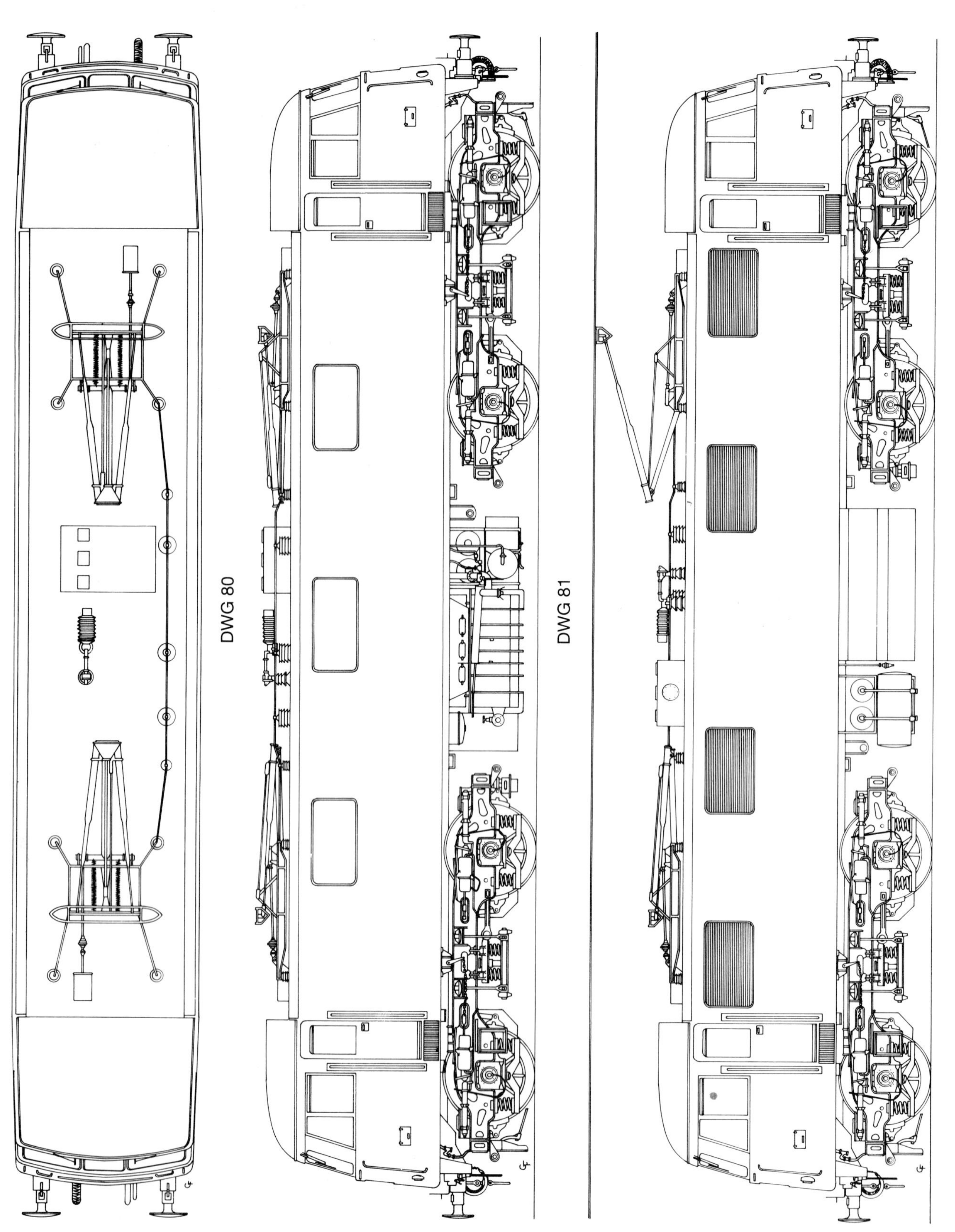

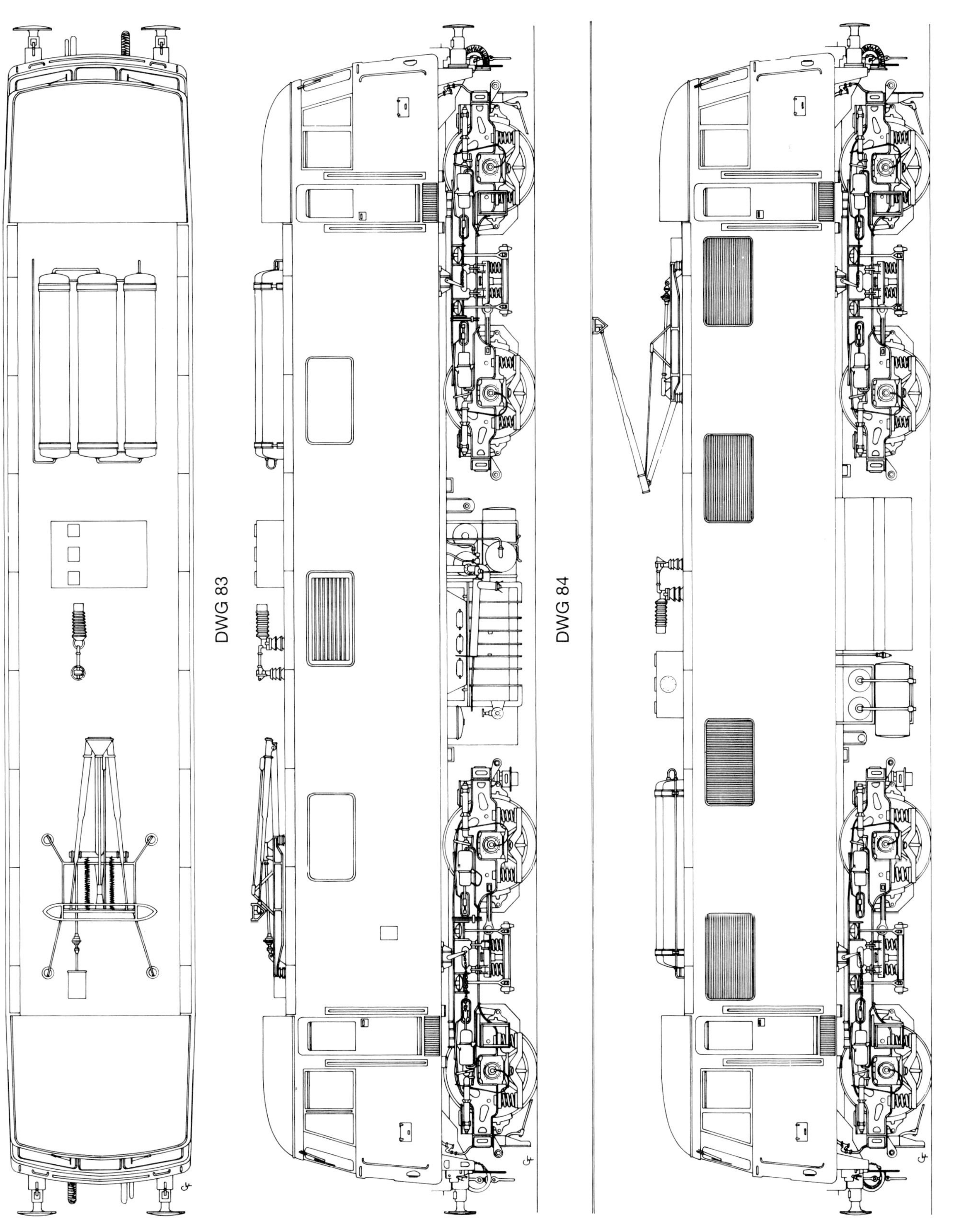
DWG 83
DWG 84
DWG 85

DWG 80
AL3/Class 83 roof detail, showing as-built condition with two pantographs.

DWG 81
AL3/Class 83 side 'B' elevation, showing original layout, with two pantographs. No. 1 end on right.

DWG 82
AL3/Class 83 side 'A' elevation, showing original layout, with two pantographs. No. 1 end on left.

DWG 83
Class 83 roof detail, after refurbishment with one pantograph. No. 1 end on right.

DWG 84
Class 83 side 'B' elevation, after refurbishment with grille in place of window in the central body position. No. 1 end on right.

DWG 85
Class 83 side 'A' elevation, after refurbishment. No. 1 end on left.

DWG 86
AL3/Class 83 front end layout, showing the as-built condition with vacuum only train brake equipment.

DWG 87
Class 83 front end layout, showing the fitting of marker lights in former route indicator box and the provision of dual brake equipment.

DWG 86

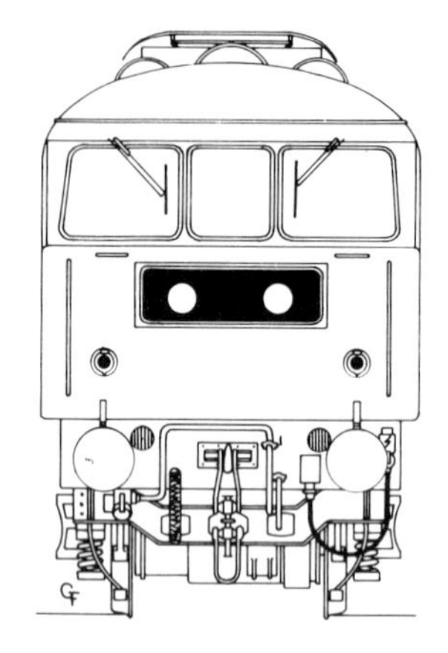

DWG 87

Class	83
Former Class Code	AL3
Number Range (TOPS)	83001-83015
Former Number Range	E3024-E3035, E3098-E3100 (E3303-E3304)
Built by	English Electric
Introduced	1960-62
Wheel Arrangement	Bo-Bo
Weight	77 tons
Height – pan down	13ft $0\frac{5}{8}$in
Width	8ft $8\frac{1}{2}$in
Length	52ft 6in
Min Curve negotiable	4 chains
Maximum Speed	100mph (Note: 1)
Wheelbase	40ft
Bogie Wheelbase	10ft
Bogie Pivot Centres	30ft
Wheel Diameter	4ft
Brake Type	Dual (Note: 2)
Sanding Equipment	Pneumatic
Heating Type	Electric - Index 66

Route Availability		6
Coupling Restriction		Not multiple fitted
Brake Force		38 tons
Horsepower	(continuous)	2,950hp
	(maximum)	4,400hp
Tractive Effort	(maximum)	38,000lb
Number of Traction Motors		4
Traction Motor Type		EE 535A
Control System		LT Tap Changing
Gear Drive		SLM flexible, single reduction
Gear Ratio		25:76
Pantograph Type		Stone-Faiveley
Rectifier Type		Silicon (Note: 3)
Nominal Supply Voltage		25kV ac

Note: 1 Locomotives remaining in service after 1986 were restricted to 40mph for ecs duties in the London area.

Note: 2 When built vacuum only braking was fitted.

Note: 3 When built mercury arc rectifiers were fitted.

The Class AL3 order consisted of 15 locomotives, originally divided as twelve Type 'A' and three Type 'B', but this was later amended to 13 Type 'A' and two Type 'B'. The BTC contract for this fleet was awarded to English Electric, who sub-contracted mechanical construction to the Vulcan Foundry at Newton-le-Willows.

The main assembly was formed of a load bearing unit, constructed from 'Corten' steel – the main underframe being an integral load bearing assembly. The actual steel skeleton frame was plated with medium gauge steel plate, while the roof areas were formed in fibre glass to keep weight to a minimum.

The internal configuration was of the standard ac type, with a walkway on one side, and locked electrical equipment compartments on the other. A problem encountered with pantograph-fitted locomotives was that of restricted roof height of the between cab section, however on the AL3 fleet this was overcome by providing a recessed walkway. Side ventilation was satisfied (on the equipment side) by four louvred panels, while the walkway side had three windows - this side was altered in later years to provide an additional ventilation grille in the former centre window position.

Vulcan Foundry commenced construction of the Type 'A' locomotives in late 1959, with the first example, No. E3024, being handed over to the BTC in July 1960, the final Type 'A' locomotive, No. E3035, entering service one year later. The Type 'B' locomotives were constructed during 1961 and allocated the numbers E3303 and E3304.

In 1960 the English Electric Co. sought permission from the BTC to construct the final locomotive with advanced power and control equipments. This permission was eventually granted, which saw the installation of a silicon rectifier, together with rheostatic brake equipment. Although under the original contract the final machine was to have been a Type 'B' freight locomotive, as the test equipment was installed, the BTC agreed to make this an additional passenger Type 'A' example, which was allocated the number E3100. To provide adequate ventilation for the new equipments this locomotive had some minor alterations to the side grille equipment.

As with previous ac types, two pantographs were originally fitted – however following the installation of dual brake equipment one pantograph, that for 6.25kV operation was removed, being replaced by additional air reservoirs.

When built the AL3 fleet were finished in the standard Electric blue, with white window surrounds and cab roof. By the late 1960s this scheme had given way to standard Rail blue with full yellow ends. Under the TOPS ac classification system the AL3s became Class 83, and renumbered in the 830xx series.

Although originally giving good service, by the late 1960s the AL3 fleet were causing the operators a number of problems – mainly involving the main power equipment. Such was the degree of difficulty that the entire class were stored in 1970. It was only the authorisation of the electrification of the WCML north of Weaver Junction to Glasgow that saved the type from being withdrawn. However with the projected shortage of motive power for the enlarged electric network, the entire fleet was refurbished.

By 1982 the fleet was again doomed, this time as 'surplus to requirements', at first the type was stored but later was withdrawn. However two examples were retained for empty stock duties in the London area, operating until 1988.

The constructional contract for the 15 strong Class AL3, later Class 83 fleet, was awarded to AEI who sub-contracted mechanical construction to English Electric, Vulcan Foundry at Newton-le-Willows. This view of the EE production line shows four locomotives under construction.

Author's Collection

As detailed in the text, two of the AL3 build were designated as Type B or freight locomotives and numbered E3303 and E3304. After a short period these two locomotives were standardised and renumbered E3098 and E3099 respectively. No. E3304 is illustrated at Allerton.

GEC Traction

Although looking the same as the production batch the final AL3, No. E3100 was constructed as an English Electric test-bed to evaluate new advances in traction equipment. After release from Vulcan Foundry No. E3100, was the subject of extensive road and performance tests for both BR and English Electric. Here it is seen with a motley train of engineering test vehicles including three further Class AL3 locomotives all under power, used to draw high current out of the overhead line equipment.

Author's Collection

After the majority of Class 83 locomotives were withdrawn from service as surplus to requirements, three machines, Nos 83009/12/15 were retained for ecs duties in the London area, being maintained by Willesden electric depot. No. 83009 is seen at Euston on 30th December 1985 with stock for the 20.30 Glasgow mail train.

Michael J. Collins

As the final Class 83 locomotives were only used for ecs movements their external condition left something to be desired, as their frequent usage through coach washing plants turned their body sides nearly white. No. 83009 sporting its 1980's-added 'X' arm pantograph, stands at Stonebridge Park on 9th April 1988.

Michael J. Collins

In their closing years the Class 83 fleet became used more frequently on secondary passenger or special duties. Here, No. 83010 is seen at Mitre Bridge Junction, Willesden after arrival with a troop special. At this point the ac electric gave way to diesel traction for the remainder of the journey to Aldershot.

Colin J. Marsden

Class 84

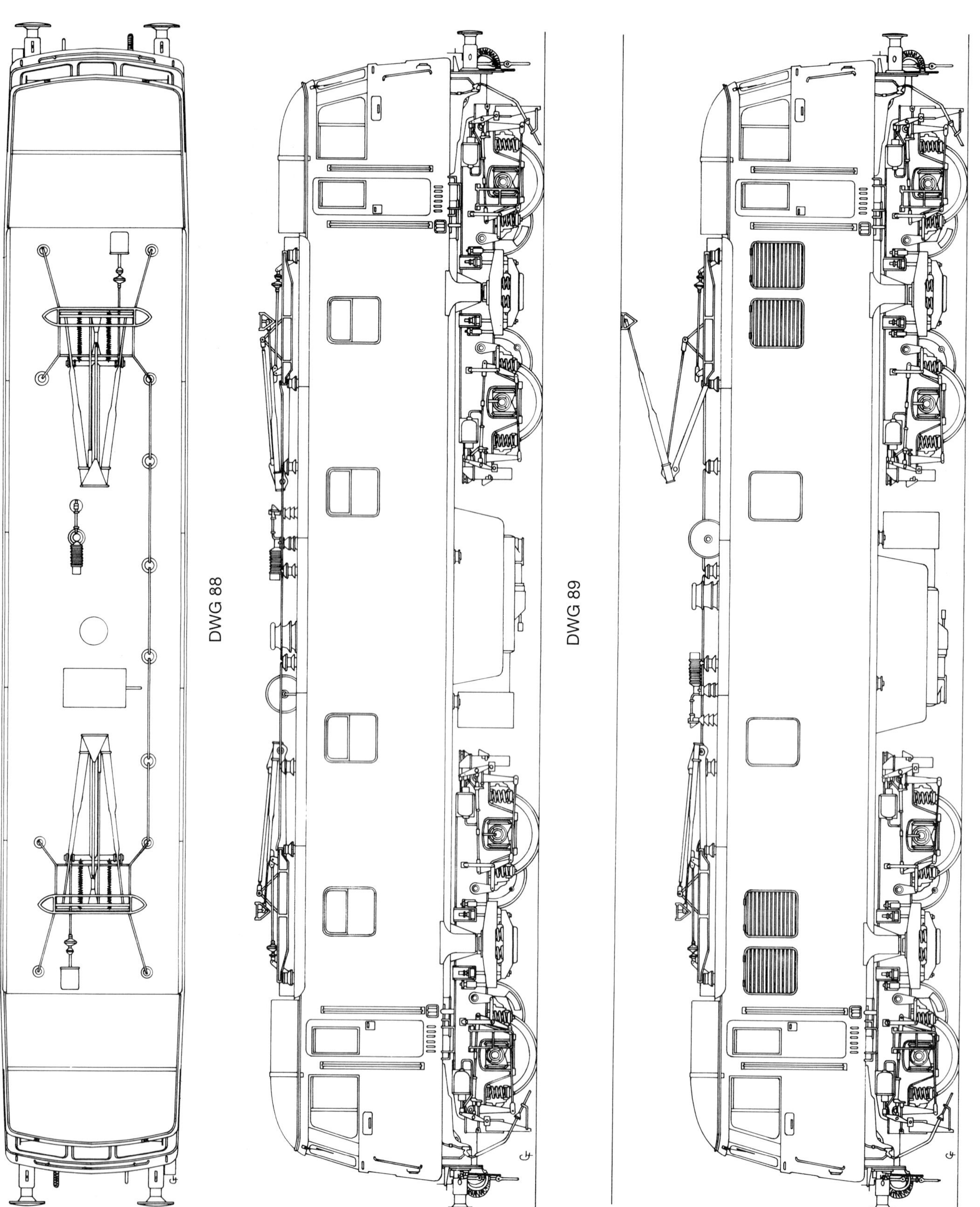

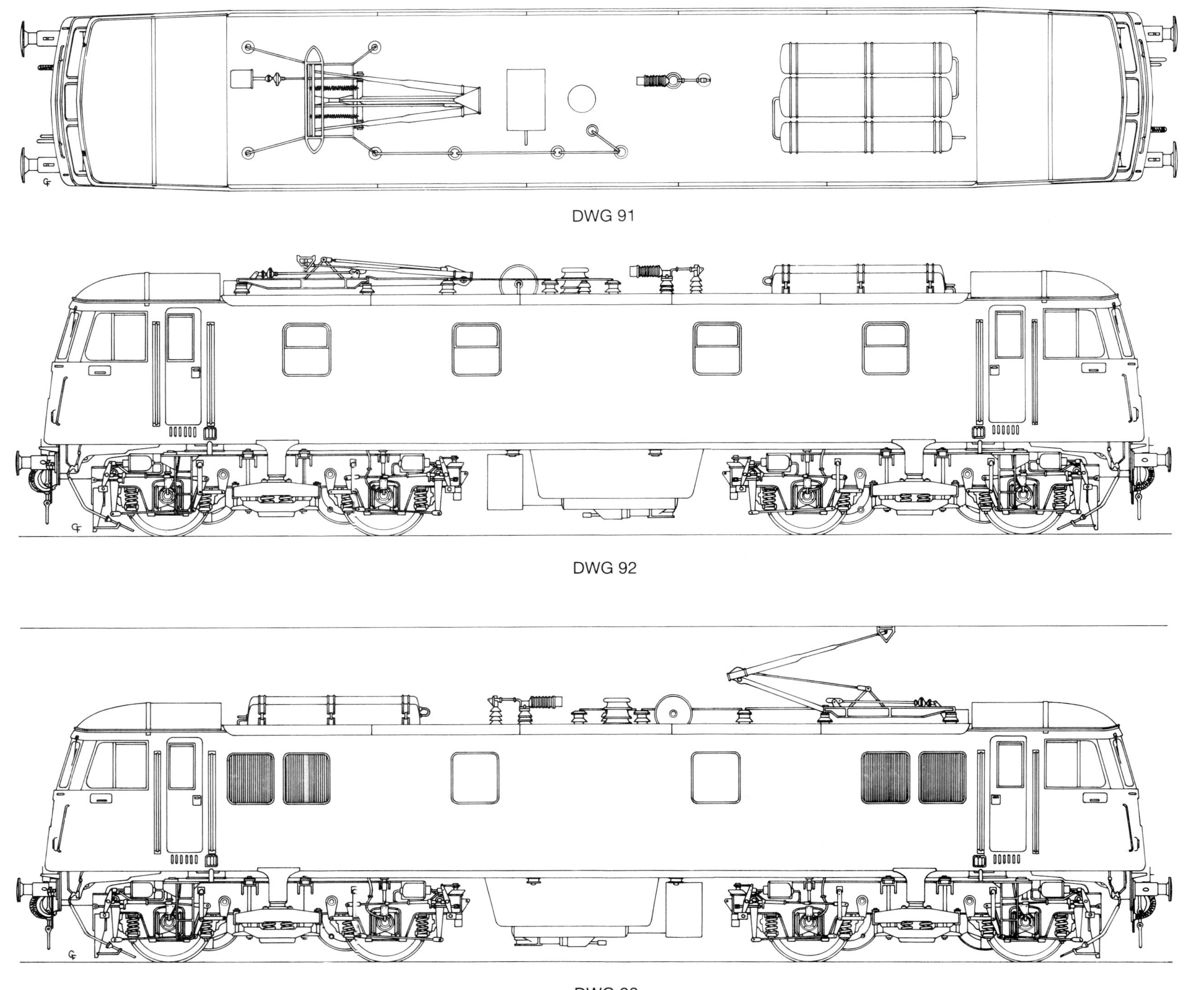
DWG 91

DWG 92

DWG 93

DWG 94

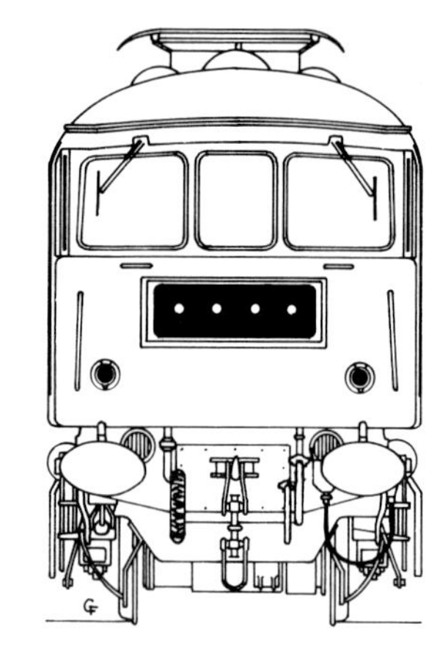

DWG 95

Class	84
Former Class Code	AL4
Number Range (TOPS)	84001-84010
Former Number Range	E3036-E3045
Built by	North British Locomotive Co. Ltd
Introduced	1960-61
Wheel Arrangement	Bo-Bo
Weight (operational)	77 tons
Height - pan down	13ft 0$\frac{5}{8}$in
Width	8ft 8$\frac{1}{4}$in
Length	53ft 6in
Min Curve negotiable	4 chains
Maximum Speed	100 mph
Wheelbase	39ft 6in
Bogie Wheelbase	10ft
Bogie Pivot Centres	29ft 6in
Wheel Diameter	4ft
Brake Type	Dual (Note:1)

Sanding Equipment		Pneumatic
Heating Type		Electric - Index 66
Route Availability		6
Coupling Restriction		Not multiple fitted
Brake Force		38 tons
Horsepower	(continuous)	3,300hp
	(maximum)	4,900hp
Tractive Effort	(maximum)	50,000lb
Number of Traction Motors		4
Traction Motor Type		GEC WT 501
Control System		HT Tap Changing
Gear Drive		Brown Boveri
Gear Ratio		25:74
Pantograph Type		Stone-Faiveley
Rectifier Type		Silicon (Note: 2)
Nominal Supply Voltage		25kV ac

Note: 1 When built vacuum only brakes were fitted.

Note: 2 When built mercury arc rectifiers were fitted.

DWG 88
AL4/Class 84 roof detail, showing the as-built condition with two pantographs. No. 1 end on left.

DWG 89
AL4/Class 84 side 'B' elevation, showing original layout.

DWG 90
AL4/Class 84 side 'A' elevation, showing original layout with one pantograph (raised).

DWG 91
Class 84 roof detail, after refurbishment with only one pantograph.

DWG 92
Class 84 side 'B' elevation, after refurbishment with only one pantograph. No. 1 end to right.

DWG 93
Class 84 side 'A' elevation, after refurbishment with only one pantograph. No. 1 end is on the left.

DWG 94
AL4/Class 84 front end layout, showing as-built condition with vacuum only train braking equipment.

DWG 95
Class 84 front end layout, showing modified design with dual brake equipment.

The ten locomotives classified by the BTC as AL4 were of Type 'A' and constructed by the North British Locomotive Co. of Glasgow, who acted as chief sub-contractor to GEC who were awarded the main contract. NBL decided to opt for an integral structure using entirely steel members, unlike other competitors who used light-weight fibre glass for some components.

The construction of this fleet commenced in mid-1959, with the first member being handed over to the BTC in March 1960. The style of the product and its livery closely followed previous types, but incorporated a slightly recessed route indicator panel, giving an immediate method of recognition. Also oval buffers were used in place of the usual round type. The between cab layout was as on previous types, with equipment on one side and a walkway on the other. The walkway side sported four aluminium framed drop light windows, the only openable equipment room windows fitted to any ac electric locomotive type. On the equipment side four grille panels and two glazed windows were provided. The number range allocated was E3036-E3045, under the later TOPS numbering system the fleet becoming Nos 84001-84010.

Following introduction and allocation to Manchester Longsight problems were soon encountered with both rough riding and failures of the main power equipment. In April 1963 the entire fleet were temporarily removed from service and sent to Dukinfield where remedial work was carried out by GEC. Regrettably, even after the class re-entered service electrical problems still ensued, which eventually led to nine of the fleet being stored at Bury from 1967, the tenth example, No. E3043 being allocated to the Rugby M&EE testing station for extensive trials.

It seemed that the AL4 fleet were doomed to be some of the shortest lived electric locomotives. However, after authorisation for the extension of the WCML electrification to Scotland there was a need for additional traction, and this requirement was fulfilled by refurbishing the Class AL4 fleet at Doncaster, a job which was completed in 1972. During the refurbishment dual brake equipment was fitted, and the redundant 6.25kV pantograph removed. At around the same time the type became classified under the TOPS numerical system as Class 84.

Many of the previous problems were overcome by the refurbishing, but others were soon identified, mainly involving the traction motor drives. Expenditure on the fleet was largely curtailed during the mid-1970s, as the BRB could not authorise any further financial investment in the class. By 1977 the first withdrawals were made, with the final member being withdrawn in 1980.

Thankfully, two members of the class were saved from scrap. No. E3044 (84009) was rebuilt as a mobile load bank for the M&EE department, and used for testing new overhead line equipment and No. E3036 (84001) which is now owned by the National Railway Museum, York. The mobile load bank, renumbered ADB968009 was withdrawn from use in Autumn 1992.

The North British built Type AL4, later Class 84, locomotives were always recognisable from their early sisters by having oval buffers, a recessed route indicator panel and drop light windows on one side. Locomotive No. E3037 is seen on display at the 1960 Institute of Transport Congress exhibition at Marylebone. Members of GEC and the BTC pose in front of the new locomotive for this official photograph.

GEC Traction

In common with the 'first' of all new types of locomotive the pioneer Class AL4, No E3036, was the subject of extensive testing once handed over to the BTC. The locomotive is seen here at the head of a rake of Mk 1 maroon coaches during a test run on the Styal line.

Author's Collection

With its distinctive North British Locomotive Co. diamond shaped builder's plate below its cast aluminium numbers, No. E3044 is seen at Manchester Piccadilly. The 'A' side of the locomotive is nearest the camera.

Norman E. Preedy

After withdrawal the majority of Class 84s were disposed of quite quickly, however No. 84008 was retained at BREL Crewe Works until 1988, although in a rather dilapidated condition. The locomotive is viewed on 25th September 1985. Note that the cast BR logo was still in position.
Colin J. Marsden

In their later years the Class 84 fleet were allocated to Crewe electric depot, but maintenance was provided by any of the LM ac electric servicing depots, only the most major repairs being returned to the owning depot. On 7th December 1974 No. 84009 poses inside Willesden depot.
Norman E. Preedy

In the months prior to their withdrawal the Class 84 fleet were deployed on several railtours. On 16th September 1978 No. 84002 was used on the 'AC/DC' tour from Birmingham to Sheffield, which travelled via the Woodhead route and also used Class 76 traction. From Birmingham to Manchester motive power was provided by No. 84002 seen here at Stockport.
Norman E. Preedy

Class 85

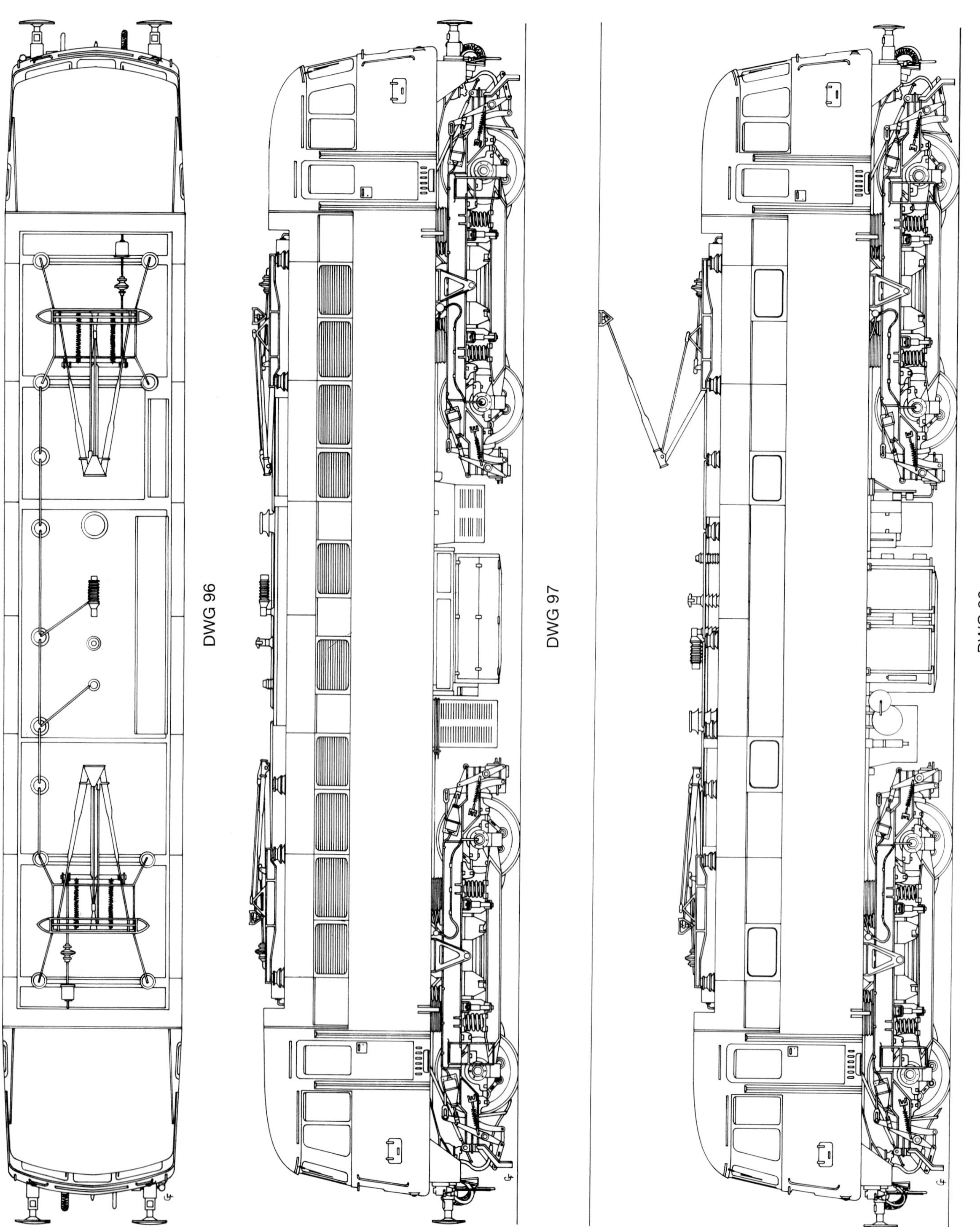

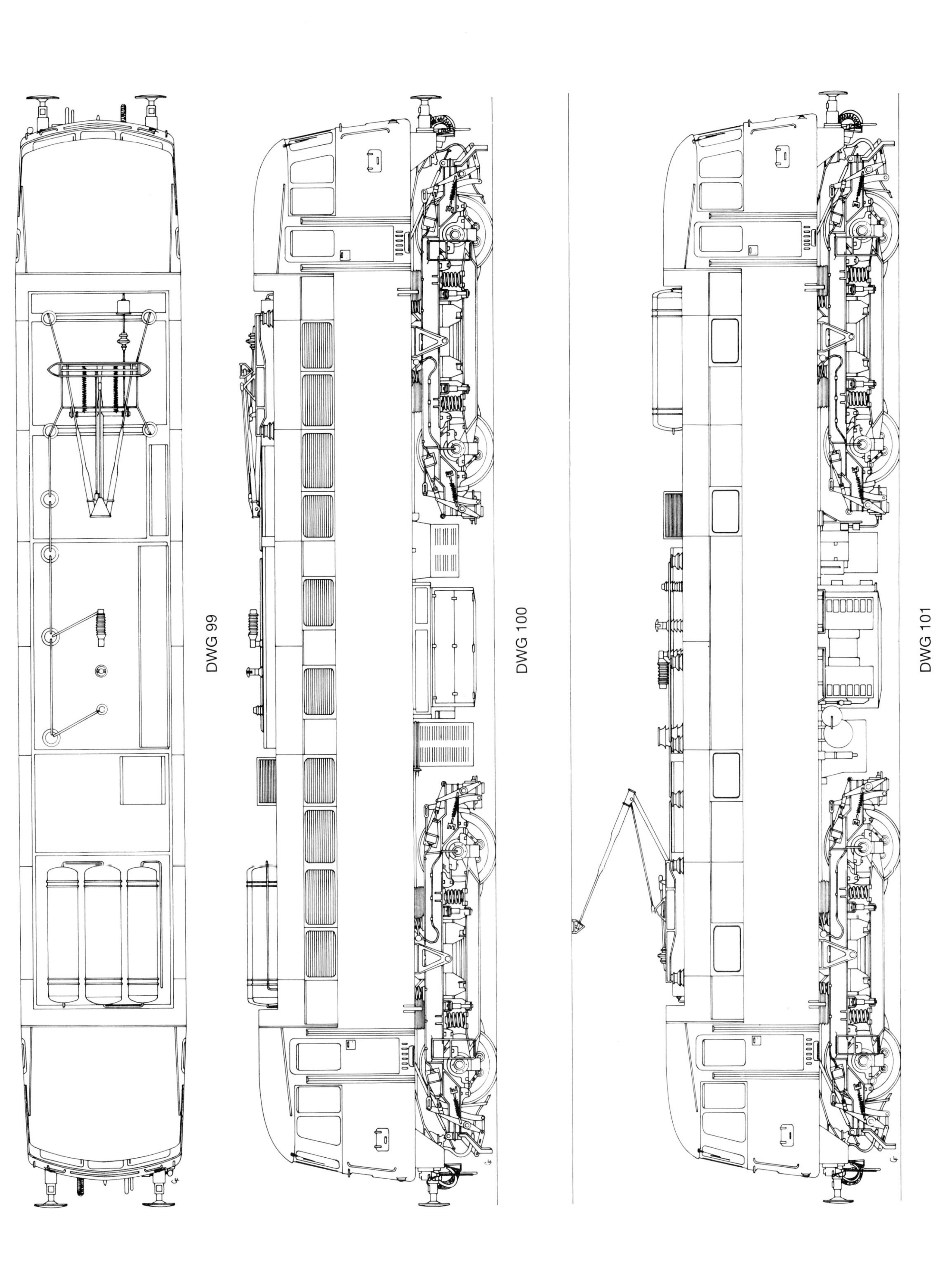
DWG 99
DWG 100
DWG 101

DWG 102

DWG 103

Class .	85
Former Class Code	AL5
Number Range (TOPS)	85001-85040, 85101-85114
Former Number Range	E3056-E3095
Built by	BR Doncaster
Introduced	1961-64
Wheel Arrangement	Bo-Bo
Weight (operational)	83 tons
Height - pan down	13ft 0$^{5}/_{8}$in
Width	8ft 8$^{1}/_{4}$in
Length	56ft 6in
Min Curve negotiable	6 chains
Maximum Speed	100mph (Note: 1)
Wheelbase	42ft 3in
Bogie Wheelbase	10ft 9in
Bogie Pivot Centres	31ft 6in
Wheel Diameter	4ft
Brake Type	Dual (Note: 2)
Sanding Equipment	Pneumatic
Heating Type	Electric - Index 66
Route Availability	6
Coupling Restriction	Not multiple fitted
Brake Force	41 tonnes
Horsepower (continuous)	3,200hp
(maximum)	5,100hp
Tractive Effort (maximum)	50,000lb
Number of Traction Motors	4
Traction Motor Type	AEI 189
Control System	LT Tap Changing
Gear Drive	Alsthom Quill, single reduction
Gear Ratio	29:76
Pantograph Type	Stone-Faiveley
Rectifier Type	Silicon (Note: 3)
Nominal Supply Voltage	25kV ac

Note: 1 From 1986 some locomotives were restricted to 80mph .

Note: 2 When built vacuum only brakes were fitted.

Note: 3 When built mercury arc rectifiers were fitted.

On Class 85/1 locomotives the ETS equipment was isolated for use on Railfreight services.

DWG 96
AL5/Class 85 roof detail, showing original layout with two pantographs.

DWG 97
AL5/Class 85 side 'A' elevation, showing original layout with two pantographs and straight style rain strips above cab door/window. No. 1 end on left.

DWG 98
AL5/Class 85 side 'B' elevation, showing original layout with two pantographs and straight style rain strips above cab door/window. No. 1 end on right.

DWG 99
Class 85 roof layout, showing refurbished locomotive with one pantograph. No 1. end on left.

DWG 100
Class 85 side 'A' elevation, showing refurbished layout, with only one pantograph, and angled rain water strip above cab door/windows. No. 1 end on left.

DWG 101
Class 85 side 'B' elevation, showing refurbished layout, with only one pantograph, and angled rain water strip above cab door/windows. No. 1 end on right.

DWG 102
AL5/Class 85 front end layout, showing original style with only vacuum train brake equipment.

DWG 103
Class 85 front end layout, showing revised design with dual brake equipment.

The main contractor for the 40 strong AL5 fleet was BR Workshops Division, who allocated the assembly work to its Doncaster Works. Power, control and technical equipment was supplied by GEC/AEI. When introduced this fleet were allocated numbers in the E3056-E3095 range, which later, under the TOPS numbering scheme, was amended to 85001-85040.

On this fleet the base underframe was formed out of seven box sections, onto which the cab ends, and lower sections of the body were assembled, forming a trough-like fabrication. The lightweight top section of the body was then mounted onto the base. One feature of this fleet was that the complete upper section between the cabs was removable and eased access for maintenance.

The internal between cab layout closely followed the previous designs, with equipment on one side and a walkway on the other. On the equipment side ten body side louvres were positioned, while the walkway side incorporated four glazed windows. When introduced two roof-mounted pantographs were installed, but in later years, after dual brake equipment was fitted the second pick-up was removed in favour of additional air reservoirs.

The Doncaster 'Plant' Works commenced production of this class in early 1960, completing the first locomotive in October, when No. E3056 was exhibited at the Electrification Conference Exhibition held at Battersea in South London. After a few early teething troubles the AL5s, later Class 85s, settled down to give good all round service, spending most of their lives allocated to Crewe Electric Depot.

After being introduced painted in standard Electric blue livery, the fleet were repainted into corporate BR Rail blue during the mid-1960s, incorporating full yellow warning ends.

The Class 85s remained in service until mid 1991, after inroads had been made into the fleet progressively from 1989. During 1989 a batch of 13 locomotives were modified by Crewe depot into freight only locomotives, having their maximum speed reduced and train heat equipment removed. These locomotives were reclassified as Class 85/1 and numbered in the series 85101-85114. Towards the later years of the locomotives' operation they became more extensively used on freight duties, mainly in the northern section of the WCML, while a small number were dedicated to empty stock movements in the London and Manchester areas.

Few major structural changes befell the class after construction, but one front end change worthy of note was the replacement of the four-character route indicator boxes with sealed beam marker lights during the early 1980s.

Of the five original 'prototype' AL classes introduced, only one fleet, the AL5s were constructed by a BR workshops – Doncaster, who employed AEI/EE as the chief sub-contractor. No. E3079, a 1963 built locomotive, is seen heading the 4.20pm Liverpool Lime Street – Euston express past Runcorn on 30th March 1965.

BR

Although all were introduced without a yellow warning panel front end, this was soon applied, and by the end of 1965 all examples had this addition. With a splendid mixture of maroon liveried LMS/BR coaching stock behind, No. E3087 pulls off the Nuneaton line at Rugby on 14th September 1966 with the 10.00 Liverpool – Euston service.

Author's Collection

During the early 1970s the entire Class AL5 fleet were refurbished by Doncaster Works, when amongst other work, dual brake equipment was installed and the dual voltage power equipment removed, leaving only one roof-mounted pantograph. No. 85025 is seen near Basford Hall, Crewe on 16th July 1985 heading the 09.10 Willesden – Warrington 'Speedlink' service.

John Tuffs

Although not diagrammed for use on crack express duties towards the end of their careers, the Class 85s were often seen at the head of InterCity rakes operating on both the London Midland and Scottish regions as late as the early 1990s. On 25th October 1981 No. 85001 passes Castlethorpe with the 09.50 Liverpool Lime Street – Euston.

Michael J. Collins

With its pantograph end leading, and the locomotive's 'A' side nearest the camera, No. 85028 passes Sandon between Stafford and Crewe with the 11.15 Euston – Manchester service on 20th June 1985.

John Tuffs

The final duties for the Class 85s were for the Railfreight sector, which took the machines to all parts of the 25kV electrified network, and indeed during 1988 members of the fleet traversed the North London Line connection to the ER with examples being recorded at Temple Mills and Stratford. On 4th June 1979 No. 85010 is seen struggling up Shap Bank at Greenholme with a 'Cartic' train bound for Scotland.

Colin J. Marsden

Class 86

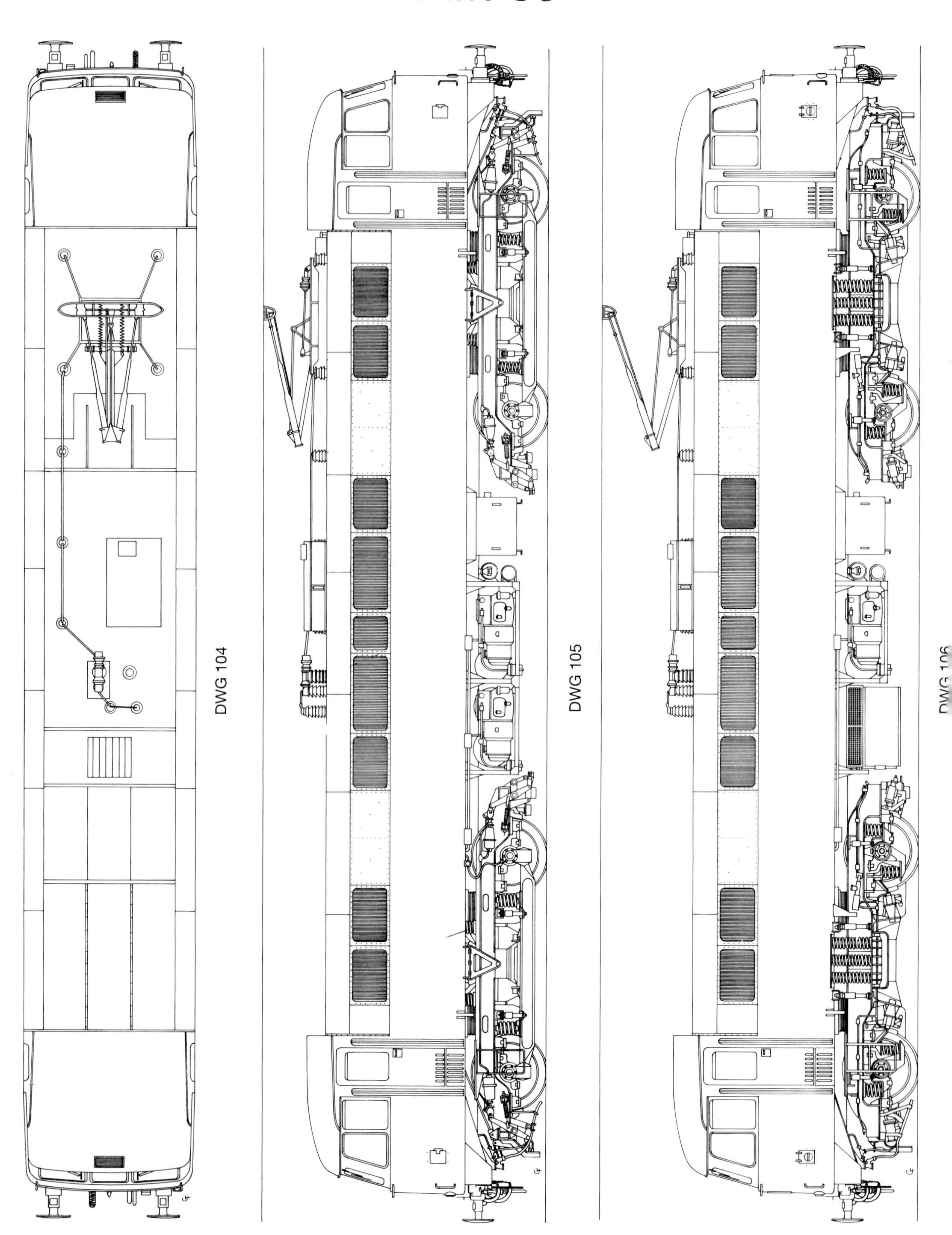

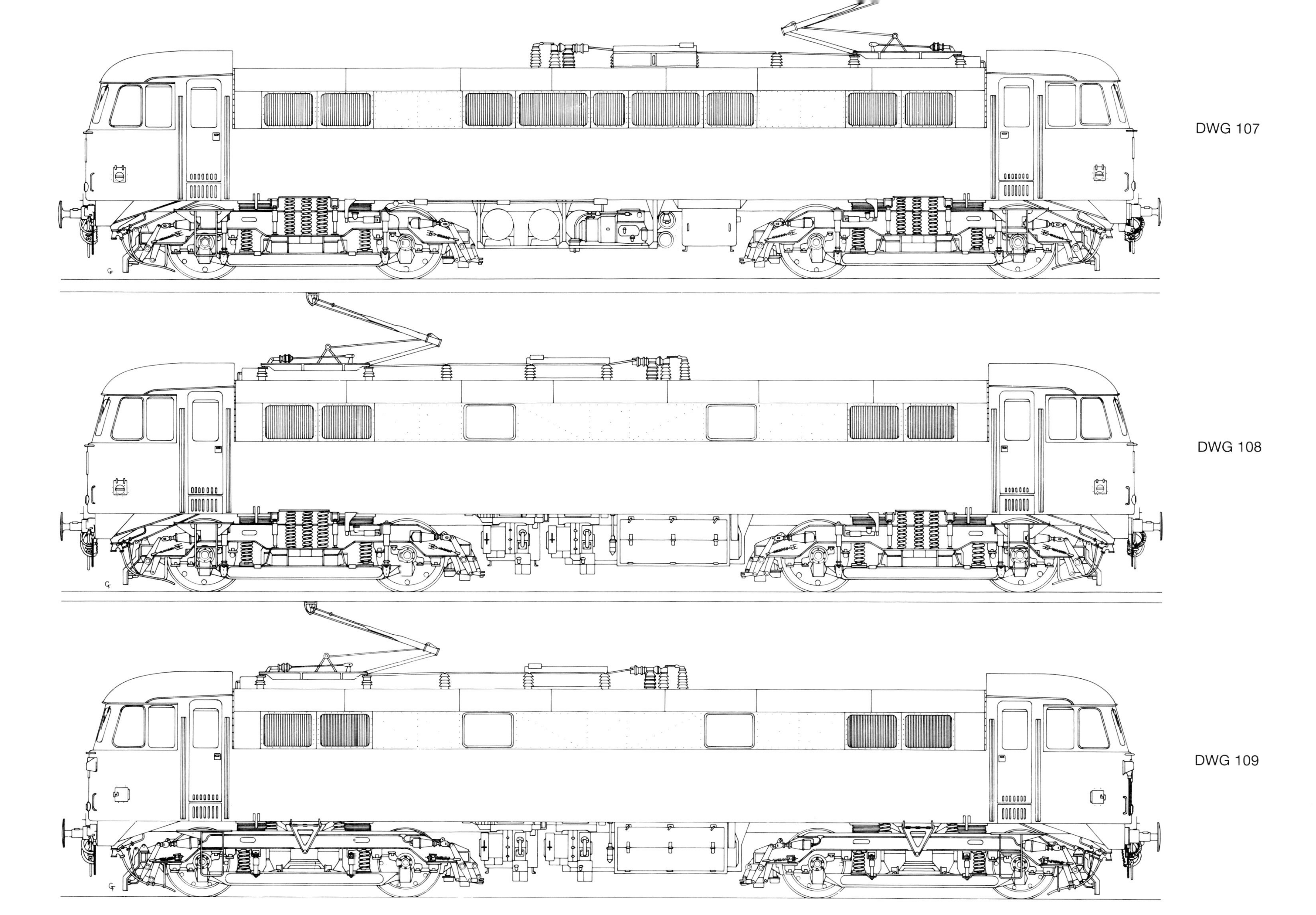
DWG 107
DWG 108
DWG 109

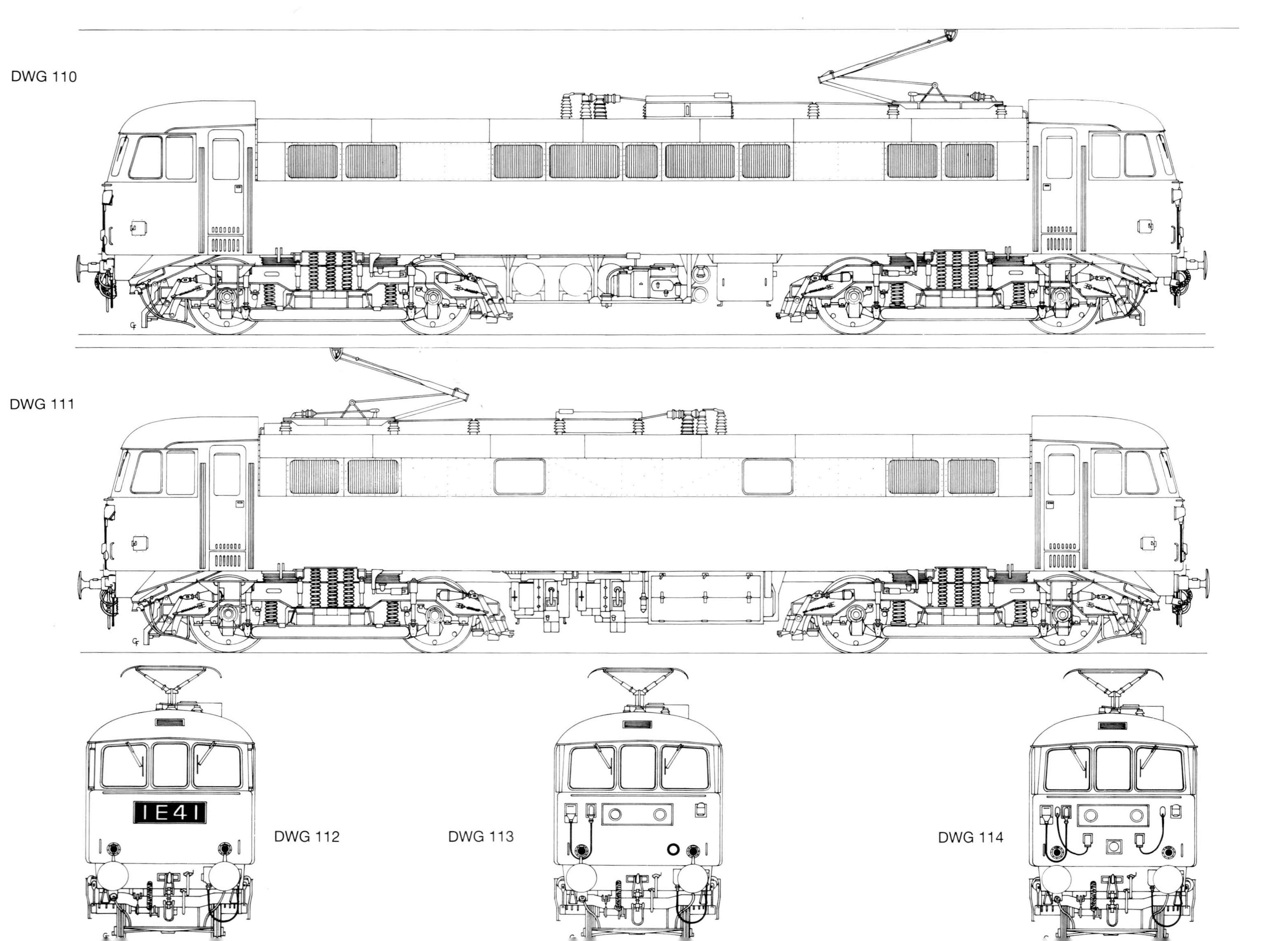
DWG 110
DWG 111
IE41
DWG 112
DWG 113
DWG 114

DWG 104
Class 86 roof detail, No. 1 end on left.

DWG 105
Class 86 side 'A' elevation, showing original as-built condition, drawing also applicable to Class 86/0. No. 1 end on left.

DWG 106
Class 86/1 side 'A' elevation, No. 1 end to left.

DWG 107
Class 86/2 side 'A' elevation, showing top hinged sand boxes. No. 1 end on left.

DWG 108
Class 86/2 side 'B' elevation, showing top hinged sand boxes. No. 1 end on right.

DWG 109
Class 86/3 side 'B' elevation, showing side hinged sand boxes. No. 1 end on right.

DWG 110
Class 86/4, 86/6 side 'A' elevation. No. 1 end on left.

DWG 111
Class 86/4 side 'B' elevation. No. 1 end on right.

DWG 112
Class 86 front end layout, showing original as built condition.

DWG 113
Class 86 front end layout, showing headcode box fitted with sealed beam marker lights, quartz headlight and multiple control jumpers.

DWG 114
Class 86 front end layout showing the style fitted with marker lights, central headlight, multiple control jumpers and Time Division Multiplex (TDM) equipment. Some locomotives now have conventional jumpers removed with blanking plates fitted.

Class	86/1	86/2	86/4	86/6
Former Class Code	AL6	AL6	AL6	AL6
Number Range	86101-86103	86204-86261	86401-86439*	86601-86639*
Former Number Range	Note: 1	Note: 1	Note: 1	Note: 1
Built by	EE Ltd	EE Ltd & BR Doncaster	EE Ltd & BR Doncaster	EE Ltd & BR Doncaster
Introduced (Note: 2)	As 86/1 1972	As 86/2 1972-75	As 86/4 1984-87	As 86/6 1990
Wheel Arrangement	Bo-Bo	Bo-Bo	Bo-Bo	Bo-Bo
Weight (operational)	87 tons	85 tons (Note: 3)	83 tons (Note: 3)	83 tons
Height - pan down	13ft 0$\frac{9}{16}$in	13ft 0$\frac{9}{16}$in	13ft 0$\frac{9}{16}$in	13ft 0$\frac{9}{16}$in
Width	8ft 8$\frac{1}{4}$in	8ft 8$\frac{1}{4}$in	8ft 8$\frac{1}{4}$in	8ft 8$\frac{1}{4}$in
Length	58ft 6in	58ft 6in	58ft 6in	58ft 6in
Min Curve negotiable	6 chains	6 chains	6 chains	6 chains
Maximum Speed	110mph	100mph (Note: 4)	100mph	75mph
Wheelbase	43ft 6in	43ft 6in	43ft 6in	43ft 6in
Bogie Wheelbase	10ft 9in	10ft 9in	10ft 9in	10ft 9in
Bogie Pivot Centres	32ft 9in	32ft 9in	32ft 9in	32ft 9in
Wheel Diameter	3ft 9$\frac{1}{4}$in	3ft 9$\frac{1}{2}$in	3ft 9$\frac{1}{2}$in	3ft 9$\frac{1}{2}$in
Brake Type (Note: 5)	Dual	Dual	Dual	Dual
Sanding Equipment	Pneumatic	Pneumatic	Pneumatic	Pneumatic
Heating Type	Electric - Index 74	Electric - Index 74	Electric - Index 74	Isolated
Route Availability	6	6	6	6
Coupling Restriction	TDM fitted	TDM fitted	TDM fitted	TDM fitted
Brake Force	40 tons	40 tons	40 tons	40 tons
Horsepower (continuous)	5,000hp	4,040hp	3,600hp	3,600hp
(maximum)	7,860hp	6,100hp	5,900hp	5,900hp
Tractive Effort (maximum)	58,000lb	46,500lb	58,000lb	58,000lb
Number of Traction Motors	4	4	4	4
Traction Motor Type	GEC G412AZ	AEI 282BZ	AEI 282AZ	AEI 282AZ
Control System	HT Tap Changing	HT Tap Changing	HT Tap Changing	HT Tap Changing
Gear Ratio	32:73	22:65	22:65	22:65
Pantograph Type	Brecknell Willis	Brecknell Willis/AEI	Stone Faiveley/AEI	Stone Faiveley/AEI
Rectifier Type	Silicon Semi Conductor	Silicon Semi Conductor	Silicon Semi Conductor	Silicon Semi Conductor
Nominal Supply Voltage	25kV AC	25kV AC	25kV AC	25kV AC

*Not consecutive numbering

Note: 1 The Class 86 original numbers were E3101-E3200, renumbering was carried out at random, as modification work was effected.

Note: 2 The Class 86s were originally introduced in 1965-66.

Note: 3 A number of Class 86/2s and 86/4s have ballast weights which increase their weight by 1 ton.

Note: 4 Class 86/2s Nos 86209/24/25/31 are fitted with Brecknell Willis high speed pantographs, thus increasing the top speed to 110mph.

Note: 5 Locomotives fitted with rheostatic brake equipment.

The AL6, or Class 86, fleet of 100 ac locomotives represents the BRB's second generation of main line electric traction. The order for the fleet was placed during 1963 for English Electric/AEI to supply power/control equipments, with mechanical construction divided between English Electric's subsidiary Vulcan Foundry and the BR workshops at Doncaster.

The basic design for the fleet was based on the previous first generation types, but much of the internal equipment was revised, both to introduce new technology and improve the internal layout. External body alterations included revision of the cab end design to incorporate a flat lower body panel, and raked back front screen panel. The equipment side of the locomotive incorporated nine air louvre panels, whilst the cab-cab walkway side was fitted with four air louvred panels and two windows.

As there was no intention to install the AL6 fleet for dual voltage operation (6.25kV and 25kV), only one pantograph was fitted from new.

A total change from previous ac locomotive practice came in the traction equipments, which on this design were of the axle hung type. At the design stage it was considered these would give improved ride characteristics over the previously used frame-mounted traction motors, but in practice this was far from the case. Unfortunately the traction motor and bogie problems later led to serious bogie frame fractures and concern over track damage. To overcome such problems Flexicoil suspension was fitted on an experimental basis from the early 1970s, and progressively to the entire fleet, as were SAB low track-force wheel sets.

In the early 1970s, when planning and design was being afforded to the next generation of ac traction, three Class 86s were rebuilt as Class 87 test-beds, being installed with Class 87 style bogies, incorporating fully spring-borne traction motors, as well as much revised electrical equipment.

When built the AL6 design was finished in Electric blue livery, which was for the first few examples, devoid of a yellow warning panel. Over the years full yellow ends were applied and the blue livery amended to the corporate scheme. After the introduction of the various new business sectors, InterCity colours in various guises appeared, with the latest swallow livery being applied from 1988. Also in 1988 one member, No. 86401, was finished in Network SouthEast livery, and another in a 'mock' 1960's Electric blue scheme.

As the sectorisation of BR spread during the late-1980s/early-1990s further livery variations emerged, including Railfreight triple grey and Rail express systems red.

The area of operation of this fleet has changed considerably over the years. When first introduced the class operated entirely on the West Coast Main Line, being allocated to Willesden. Their operating range first increased in the mid-1970s following electrification of the route north of Crewe to Glasgow. During the mid-1980s after the East Anglia electrification was complete, the Class 86s were deployed on Liverpool St – Norwich duties at first allocated to Ilford and later to Norwich. Following the electrification of the East Coast Main Line from King's Cross, some members of the fleet have been deployed on this line. Today a batch are allocated to Manchester for InterCity Cross Country duties.

Following various modification programmes, four subclasses now exist within the Class 86 fleet: 86/1 – Class 87 test-bed locomotives, 86/2 – general passenger fleet fitted with Flexicoil suspension and GEC 282BZ traction motors, 86/4 – locomotives operated by the Parcels business and fitted with Flexicoil suspension and GEC 282AZ traction motors and 86/6 – locomotives used by the Freight business, restricted to 60mph, with the electric train supply equipment isolated.

In common with many of the main line classes headlights have been fitted in more recent years, while the redundant four-character route indicator boxes have been plated over or removed, being replaced by sealed beam marker lights. During the mid-1980s Time Division Multiplex (TDM) jumpers of the RCH style have been fitted to most examples of the fleet, as were multiple control jumpers to the Class 86/4 and 86/6 fleet which by mid-1992 started to be removed as the TDM system became more satisfactory.

After the resurrection of BR's naming policy in the 1970s the larger proportion of the Class 86 fleet now have names, mainly of the cast standard style.

Since their introduction only two members of the fleet have been withdrawn, after receiving collision damage. It is planned that this fleet will remain in traffic for many years to come.

Following the decision to order a large 'production' batch of 25kV locomotives for LM operation in 1964, constructional contracts were placed with both English Electric/Vulcan Foundry and the BR workshops at Doncaster to build a total of 100 locomotives. This view shows the main production line at Vulcan Foundry, with six locomotives in various stages of assembly.
Colin J. Marsden

When the first batch of Class AL6 locomotives were released from Vulcan Foundry a number of major tests were carried out, including an investigation into wheel slip/slide problems as, when built, no such automatic correction equipment was fitted. No. E3161, without a yellow warning panel front end stands at Rugby in Summer 1965 during wheel slip/slide tests. Note the water hose pipes along the solebar and front of the locomotive used to induce wheel slip/slide.

GEC Traction

By the time the Class AL6 locomotives had been introduced into regular service the small yellow warning end had been adopted as standard, and applied to all new locomotives after August 1965. No. E3146 heads a long unfitted freight through Bletchley on 18th June 1966.

John Faulkner

With its No. 2 or pantograph end leading, No. E3193 passes Wembley station during early 1966 with a Euston – Manchester express service.

GEC Traction

Clearly displaying the underframe and bufferbeam equipment layout, No. E3136 is seen at the head of a Ford Motor Company car train bound for Halewood from the Ford works at Dagenham. This locomotive was one of the BR Doncaster built examples.

Colin J. Marsden

Another train operated by the Ford Motor Company between Dagenham and Halewood was a daily parts service, which was diesel hauled between Dagenham and Willesden, and electrically hauled forward. On 14th February 1967 No. E3185 slowly departs from Willesden with the northbound service.

Author's Collection

After the introduction of new electric operated Anglo-Scottish "Electric Scot" services from 1974 the Class 86 locomotives were used alongside the newly introduced Class 87s on the new fast InterCity services. On 9th February 1983 Class 86/2 No. 86257 *Snowdon* passes Grayrigg with the 07.23 Glasgow/07.06 Edinburgh – Birmingham service.

Colin J. Marsden

Following the abolition of four-character route indicator displays during the mid-1970s the Class 86s, in common with other types, had the equipment plated over, with two marker lights fitted in the former indicator position. On 27th March 1982 Class 86/2 No. 86251 *The Birmingham Post* approaches Nuneaton with a Manchester – Euston service.

Colin J. Marsden

After the electrification of the East Anglian lines to Ipswich and onto Norwich the Class 86s became stable power for these accelerated services. Sporting full InterCity livery, No. 86220 *The Round Tabler* stands at Colchester on 5th April 1990 with a Liverpool Street – Norwich working.

Colin J. Marsden

With the introduction of InterCity and sector liveries the Class 86 fleet have all been re-painted into new revised colour schemes, the majority now displaying InterCity livery of various guises, while others carry Railfreight Distribution and Res colours. On 20th March 1988 No. 86259 *Peter Pan* passes the site of the former Ipswich steam depot with the 10.55 Norwich – Liverpool Street service.

Michael J. Collins

Class 86/4 No. 86428 *Aldaniti* was the subject of windscreen wiper research during the mid-1980s, when the wiper unit on the assistant driver's side at No. 2 end was repositioned to give an up/down wipe of the window. With modified end nearest the camera, No. 86428 passes Greenholme on 16th July 1986 with the 10.40 Glasgow – Birmingham service.

Colin J. Marsden

Class 86/2 No. 86235 *Novelty*, crosses the River Colne at Stanway Viaduct near Colchester on 16th February 1988 with the 11.30 Liverpool Street – Norwich service. From 1988 the Class 86 fleet were also introduced on selected freight diagrams in East Anglia.

Michael J. Collins

With BR's interest in improving its public image, naming of selected locomotive types was re-introduced in the 1970s, with the Class 86 fleet some of the first to benefit. On 22nd January 1985 No. 86226 was named *Royal Mail Midlands* in a special ceremony at Birmingham Moor Street.

Colin J. Marsden

The first example of the Class 86/4 fleet, No. 86401 was painted in Network SouthEast livery during 1986, and was intended for use on that sector's traffic. However on 22nd July 1987 the locomotive was seen departing from Mossend with the 20.25 freight service to Bescot. This locomotive is now painted in Rail express systems colours.

Maxwell H. Fowler

Class 87

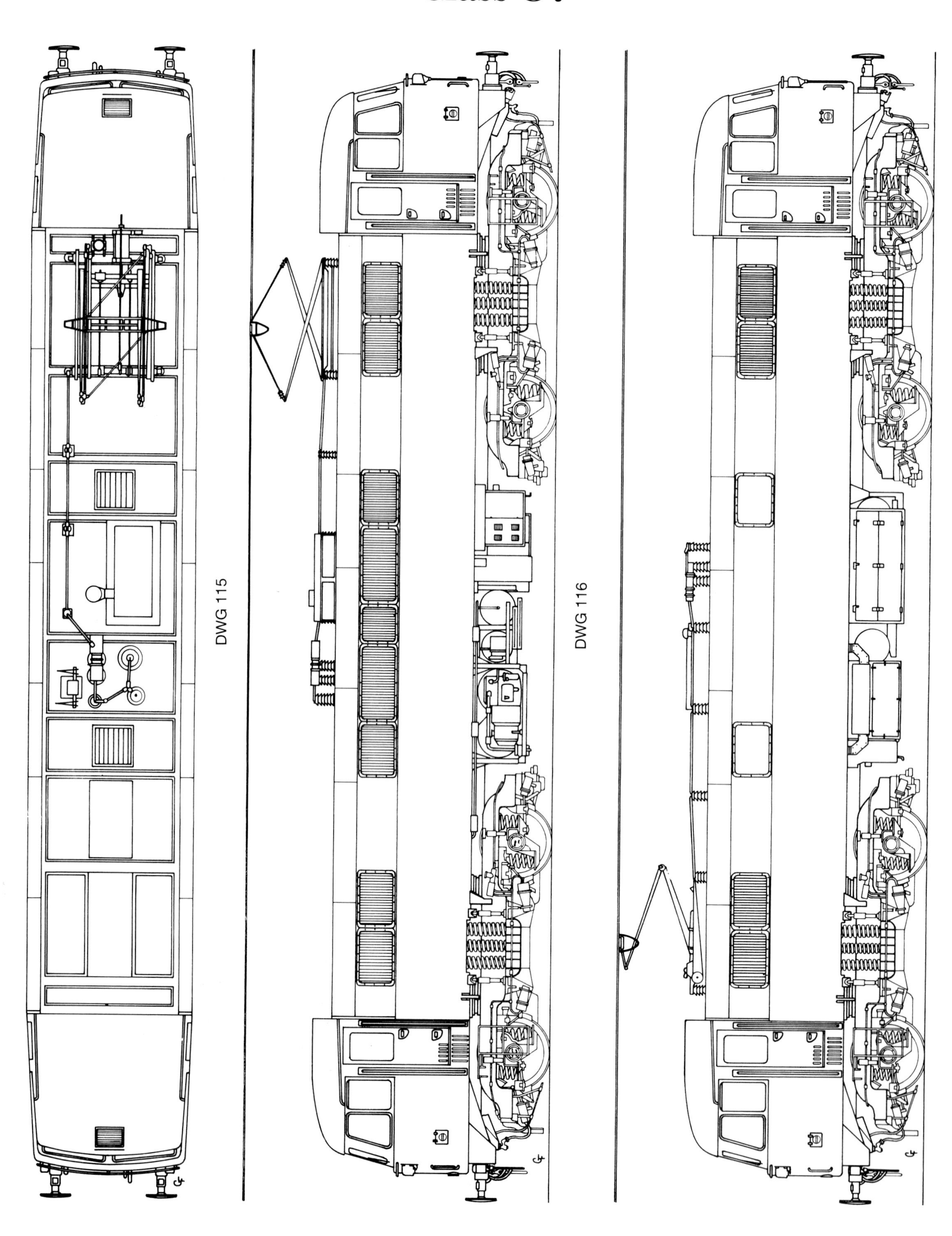

DWG 118

DWG 115
Class 87 roof detail, fitted with original cross arm pantograph. No. 1 end on left.

DWG 116
Class 87 side 'A' elevation, showing original cross arm pantograph. No. 1 end on left.

DWG 117
Class 87 side 'B' elevation, showing the Brecknell Willis High Speed pantograph. No. 1 end on right.

DWG 118
Class 87 front end layout, showing original layout.

DWG 119
Class 87 front end layout, showing Time Division Multiplex (TDM) jumpers.

DWG 119A
Class 87 front end layout, showing the removal of the jumper cable equipment.

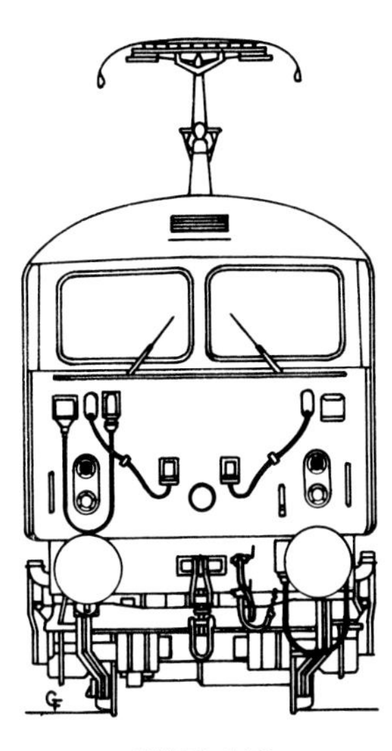

DWG 119

Class	87/0	87/1
Number Range	87001-87035	87101
Built by	BREL Crewe	BREL Crewe
Introduced	1973-74	1977
Wheel Arrangement	Bo-Bo	Bo-Bo
Weight	83 tons	79 tons
Height - pan down	13ft 1¹/₄in	13ft 1¹/₄in
Width	8ft 8¹/₄in	8ft 8¹/₄in
Length	58ft 6in	58ft 6in
Min Curve negotiable	4 chains	4 chains
Maximum Speed	110mph	75mph
Wheelbase	43ft 6¹/₈in	43ft 6¹/₈in
Bogie Wheelbase	10ft 9¹/₈in	10ft 9¹/₈ in
Bogie Pivot Centres	32ft 9in	32ft 9in
Wheel Diameter	3ft 9¹/₂in	3ft 9¹/₂in
Brake Type	Air	Air
Sanding Equipment	Pneumatic	Pneumatic
Heating Type	Electric - Index 95	Isolated
Route Availability	6	6
Coupling Restriction	TDM fitted	TDM fitted
Brake Force	40 tons	40 tons
Horsepower (continuous)	5,000hp	4,850hp
(maximum)	7,860hp	7,250hp
Tractive Effort (maximum)	58,000lb	58,000lb
Number of Traction Motors	4	4
Traction Motor Type	GEC G412AZ	GEC G412BZ
Control System	HT Tap Changing	Thyristor
Gear Ratio	32:73	32:73
Pantograph Type	Brecknell Willis HS	Brecknell Willis HS
Nominal Supply Voltage	25kV ac	25kV ac

DWG 119A

The Class 87 fleet were ordered following the authorisation for electrification north from Weaver Junction to Glasgow, as the existing fleets of ac electric locomotives, dating from the late 1950s and 1960s, would be insufficient for the new scheme. A batch of 36 locomotives were therefore ordered, classified Class 87, and numbered in the TOPS 87xxx series.

Power and control equipment for the fleet was supplied by GEC Traction, while mechanical construction was effected by BREL Crewe Works. The physical appearance of the locomotives closely resembled the previous types, but they had only two front windscreens in place of the time honoured three. Sealed beam headlights and marker lights were also installed as front indication. A major departure from previous designs was the use of Flexicoil suspension and frame-mounted traction motors.

The mechanical portion consisted of a fabricated underframe onto which the body was assembled, the upper section of the body and roof being arranged as a removable unit to assist with maintenance. The two body sides were of completely different designs, one having a near complete bank of air louvres, while the other housed two glazed windows and four louvred panels. With the adaption of modern technology the Class 87 fleet were built from new with air brake only equipment. When first introduced this did preclude the operators from obtaining the maximum availability from the class as much vacuum braked stock was still in service. The Class 87s are all allocated to Willesden except No. 87101 which is at Crewe for freight operations. The locomotives at Willesden are operated by InterCity on WCML express duties.

Power collection for the Class 87 fleet was originally by one GEC cross-arm pantograph, however by the mid-1980s these were replaced by Brecknell Willis high speed units permitting speeds of up to 110mph to be authorised.

When introduced in 1973/4 the fleet were painted in conventional all-over Rail blue livery, with full yellow ends, this remaining until 1984 when various livery experiments were carried out, resulting in the InterCity livery we see today. Soon after the fleet was introduced the BR naming policy was revived and the Class 87s became some of the first recipients. From their introduction the fleet have been allocated to Willesden from where their main responsibility has been Anglo-Scottish and North Western services.

The final locomotive of the build, allocated No. 87036, was constructed by GEC/BR as a test-bed for the use of thyristors in traction control systems. Due to the many differences on this locomotive, it was decided to number the machine 87101 to avoid confusion with the conventional locomotives. After its release to service No. 87101 was the subject of extensive testing to ascertain the benefits of the installation of modern electronics in the traction system. Much of the data obtained paved the way for the application of GEC thyristor control principles in Class 90 and 91 locomotives.

Very few structural alterations have taken place since this fleet was introduced, the most noticeable being the installation of RCH style nose end jumpers now used for TDM control, and the removal of the original multiple unit control boxes.

Following the introduction of the Class 87 electric locomotives, designed principally for Anglo-Scottish services, the majority of long-distance LM/ScR passenger duties have been so hauled. After heavy overnight snow, No. 87013 *John O' Gaunt* heads the 09.37 Carlisle – Euston past Grayrigg on 9th February 1983.

Colin J. Marsden

When introduced the Class 87 locomotives were fitted with a GEC 'X' arm pantograph, however following the decision to increase the LM line speeds over selected routes to 110mph the pantographs have been replaced by the Brecknell Willis High Speed type. Still with an 'X' arm pantograph No. 87017 *Iron Duke* passes Marston Green on 21st October 1984 with the 11.18 Wolverhampton – Euston service.
Michael J. Collins

On a number of occasions members of the Class 87 fleet have been used to haul the Royal Train, this including the Royal inaugural special when the new "Electric Scot" service was introduced. On 20th June 1977 No. 87004 (later named *Britannia*) sets off from Carnforth with stock off a Royal Train working.
Martin Welch

The Class 87 fleet were some of the first locomotives to benefit from the revised naming policy of the mid-1970s, and indeed some locomotives have been renamed over the ensuing years. The former *Redgauntlet*, No. 87026, was renamed *Sir Richard Arkwright* in a special ceremony at Preston on 12th October 1982. In this view the locomotive is seen awaiting the special unveiling ceremony.
John Tuffs

From their introduction until 1984 the Class 87 fleet were painted in conventional rail blue livery, however with the introduction of new operating businesses separate livery identities were introduced. In connection with these new liveries Class 87s Nos 87006 and 87012 were repainted in experimental schemes, No. 87006 *City of Glasgow* in all over grey, and No. 87012 *Coeur de Lion* in what is now known as InterCity livery. Both locomotives were repainted by Willesden depot where these illustrations were taken on 11th May 1984.

Both: Colin J. Marsden

Before the introduction of CEM maintenance the only location to carry out major repairs to the Class 87 fleet was BREL Crewe Works, However today, most of the classified overhauls are performed by Glasgow Springburn Works, which follows a short period when overhauls were undertaken at Stratford Major Depot. On 19th February 1991 No. 87032 *Kenilworth* poses inside Springburn Works during a classified repair.

Colin J. Marsden

When painted in their distinctive InterCity livery, hauling a rake of similarly liveried stock the Class 87s look extremely smart. Sporting TDM and multiple control jumpers, No. 87020, named *North Briton*, passes Clifton, near Penrith on 29th March 1990 with the 15.15 Edinburgh – Paddington.

Colin J. Marsden

The final member of the Class 87 build, classified as 87/1 and allocated the number 87101, was the testbed for advanced state-of-the-art traction equipments being developed during the 1970s. The machine is seen here coupled to a M&EE test car wired for a test programme in June 1975.

BR

In an early guise of InterCity livery, with its nameplate bi-secting the white body side band, No. 87006 *City of Glasgow*, departs from Euston on 25th September 1985 with the 09.45 Euston – Glasgow service.

Colin J. Marsden

During the period of livery transition, which was a protracted affair, formations of mixed livery coaching stock were a regular sight. InterCity No. 87024 *Lord of the Isles* is seen approaching Crewe on 12th February 1988 with a mixture of InterCity and blue/grey liveried vehicles forming the 13.30 Lancaster – Euston.

Colin J. Marsden

Members of the Class 87 fleet could still be seen sporting the original all-blue livery as late as Autumn 1988, this livery looking even more out of place on locomotives fitted with RCH type front jumpers. On 12th February 1988 No. 87034 *William Shakespeare* arrives at Crewe with the 07.45 Euston – Glasgow service.

Colin J. Marsden

During the late 1970s locomotive No. 87001 *Royal Scot* had its number panel applied lower on the cab side than normal, a feature shown in this illustration of the locomotive passing Blackrigg near Carlisle, on 8th June 1979, while in charge of the 07.45 Euston – Glasgow service.

Colin J. Marsden

InterCity liveried No. 87023 *Velocity* passes Crewe on 7th January 1992 with the 12.10 Liverpool Lime Street – Euston service. On the left is the rear of a northbound Liverpool train showing the MkIII Driving Van Trailer (DVT).

Colin J. Marsden

Class 89

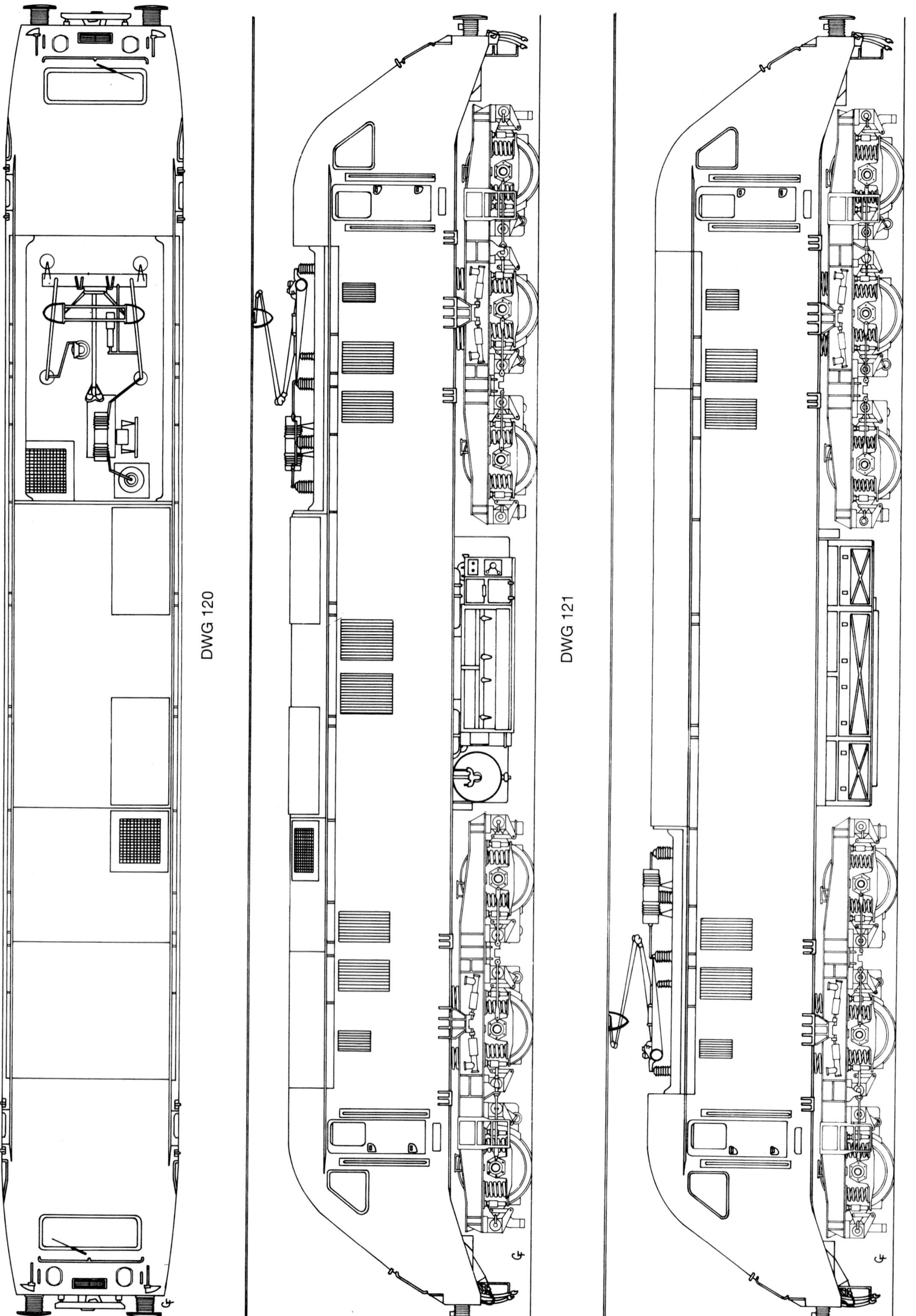

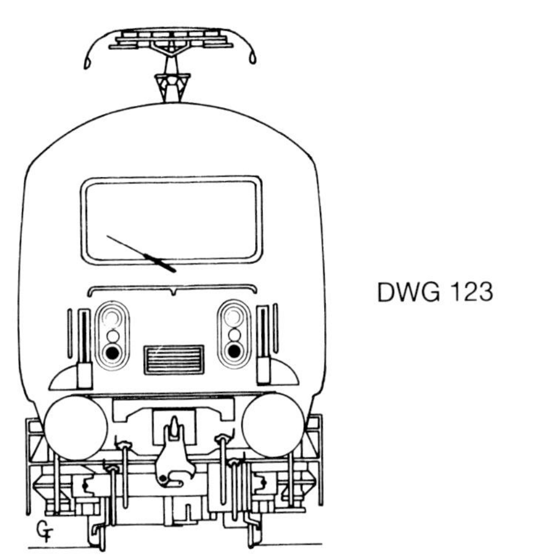

DWG 123

DWG 120
Class 89 roof detail, No. 1 end on right. Drawing shows original buffer design.

DWG 121
Class 89 side 'B' elevation. No. 1 end on right.

DWG 122
Class 89 side 'A' elevation. No. 1 end on left.

DWG 123
Class 89 front end detail.

Class	89
Number	89001
Built by	BREL Crewe & Brush Ltd
Introduced	1987
Wheel Arrangement	Co-Co
Weight	105 tons
Height – pan down	13ft 0½in
Width	8ft 11½in
Length	64ft 11in
Min Curve negotiable	4 chains
Maximum Speed	110mph*
Wheelbase	49ft 6½in
Bogie Wheelbase	14ft 5in
Bogie Pivot Centres	35ft 9¼in
Wheel Diameter	3ft 6½in
Brake Type	Air
Heating Type	Electric - Index 95
Route Availability	6
Coupling Restriction	TDM fitted
Brake Force	50 tons
Horsepower (continuous)	5,850hp
(maximum)	7,860hp
Tractive Effort (maximum)	46,100lb
Traction Motor Type	Brush TM 2201A
Control System	Thyristor
Pantograph Type	Brecknell Willis HS
Nominal Supply Voltage	25kV ac

*Designed for 125mph running.

The first electric locomotive to take to the road with a different basic design to the 1960 BTC modernisation fleet was the prototype Class 89. This class of just one locomotive, No. 89001, was constructed by BREL Crewe Works in 1985-87, as a sub-contractor to Brush, the locomotive incorporating advanced traction equipments supplied by Brush Electrical Machines of Loughbrough.

The design of the Class 89 was a complete break from any previous 25kV ac electric type, being mounted on a Co-Co wheel arrangement and incorporating streamlined body ends, reminiscent of the prototype Class 41 HST power cars. The locomotive was constructed as a 'production' demonstrator with a rating of 5,850hp, which at the time of its introduction, made it the most powerful electric locomotive in this country. Whilst its design was primarily for the high speed (200km/h) passenger market, the machine had good operating characteristics for slower speed freight duties. In having the Co-Co bogie arrangement the locomotive provided 50 per cent better tractive effort than rival Bo-Bo designs. This tractive effort would have eliminated the need for double heading of freight services over arduous inclines. However BR engineers still favoured the Bo-Bo configuration, mainly for dynamic track force reasons.

After construction by BREL Crewe the Class 89 was dispatched by rail to Derby, from where it was moved by road to Brush for exhaustive electrical testing to be carried out. After return to BR tracks the locomotive went to BREL Crewe where several minor modifications were carried out. By mid-1987 No. 89001 was transferred to the Derby-based Engineering Development Unit (EDU) where many pantograph, electrical and structural tests took place, as well as the usual 'type tests' which are carried out on all new designs. In the case of the Class 89 these were effected on the Old Dalby test track. Traction here, as the line is not electrified was provided by a Class 47 diesel propelling the locomotive and test coaches to evaluate the high speed pantograph. After acceptance by BR for main line trials the locomotive was temporarily allocated to Crewe Electric Depot, from where trials, both north over Shap and southwards to Willesden, were conducted. The performance of the locomotive soon proved to be highly successful.

By October 1987 No. 89001 had clocked up some 10,000 miles of trial running, many of which were at the head of the BREL International demonstration train on the WCML. However, due to gauge restrictions the locomotive was not permitted to enter Euston. From the end of 1987 the locomotive was transferred to the East Coast Main Line (ECML) being allocated to Hornsey and later Bounds Green, from where driver training was conducted in readiness for ECML electric services. No. 89001 was used on the ECML until May 1988 following the delivery of the first Class 91. In late May the locomotive was sent to Derby for preparation prior to being sent together with a Class 90 and 91 to Hamburg for exhibition purposes. After return to England No. 89001 resumed ECML operation, working alongside the Class 91 fleet until 1990, when due to technical defects, the locomotive was taken out of service and stored at Bounds Green. No. 89001 was finally withdrawn by BR in July 1992, and has now been preserved at the Midland Railway Centre, Butterley.

The cab layout of the Class 89 was totally different from any previous ac types, with a deep wrap round desk incorporating easy-to observe dials, and thoughtfully positioned controls. One novel feature included on the Class 89 was a speed selector switch, whereby the driver could pre-select a required road speed, open the power controller to the full position and the locomotive's electronics would do the rest, regulating the speed to the required figure. This feature worked in both acceleration and deceleration modes. Speed selection equipment was subsequently installed on both the Class 90 and 91 locomotives.

The Class 89 was fitted from new with conventional Electric train supply (ETS) equipment, buck-eye couplers (for the first time on an ac locomotive), and air and rheostatic (locomotive) braking.

When built, No. 89001 was finished in an early InterCity livery, which was later amended to the 1989 scheme. During early 1989 No. 89001 was named *Avocet* after the Royal Society for the Protection of Birds, by the then Prime Minister Margaret Thatcher, at King's Cross.

After being the subject of much trial and performance tests the Brush Class 89 was used by BR as part of its exhibit at the 1988 International Rail Exhibition in Hamburg. No. 89001, together with Class 90 No. 90008 and Class 91 No. 91003, stand in the yard of the Railway Technical Centre while being prepared for their visit to Germany.

Colin J. Marsden

Following the return of the Class 89 from Hamburg the locomotive was returned to Bounds Green, from where it commenced operation on the ECML, working with a converted DVT and IC125 formation. One of its regular duties was the 07.18 Peterborough – King's Cross, seen here on 30th August 1988 passing Arlesey near Biggleswade.

Brian Morrison

Whilst working on the ECML the regular northbound commuter service between King's Cross and Peterborough was the 17.36 departure from London, seen here on 10th August 1988 near Welham Green.

Brian Morrison

When the Class 89 emerged from BREL Crewe in 1986 it was fitted with a new design cab layout, not previously seen on a British electric locomotive. This illustration shows the general layout of the No. 2 end cab.

Michael J. Collins

Class 90

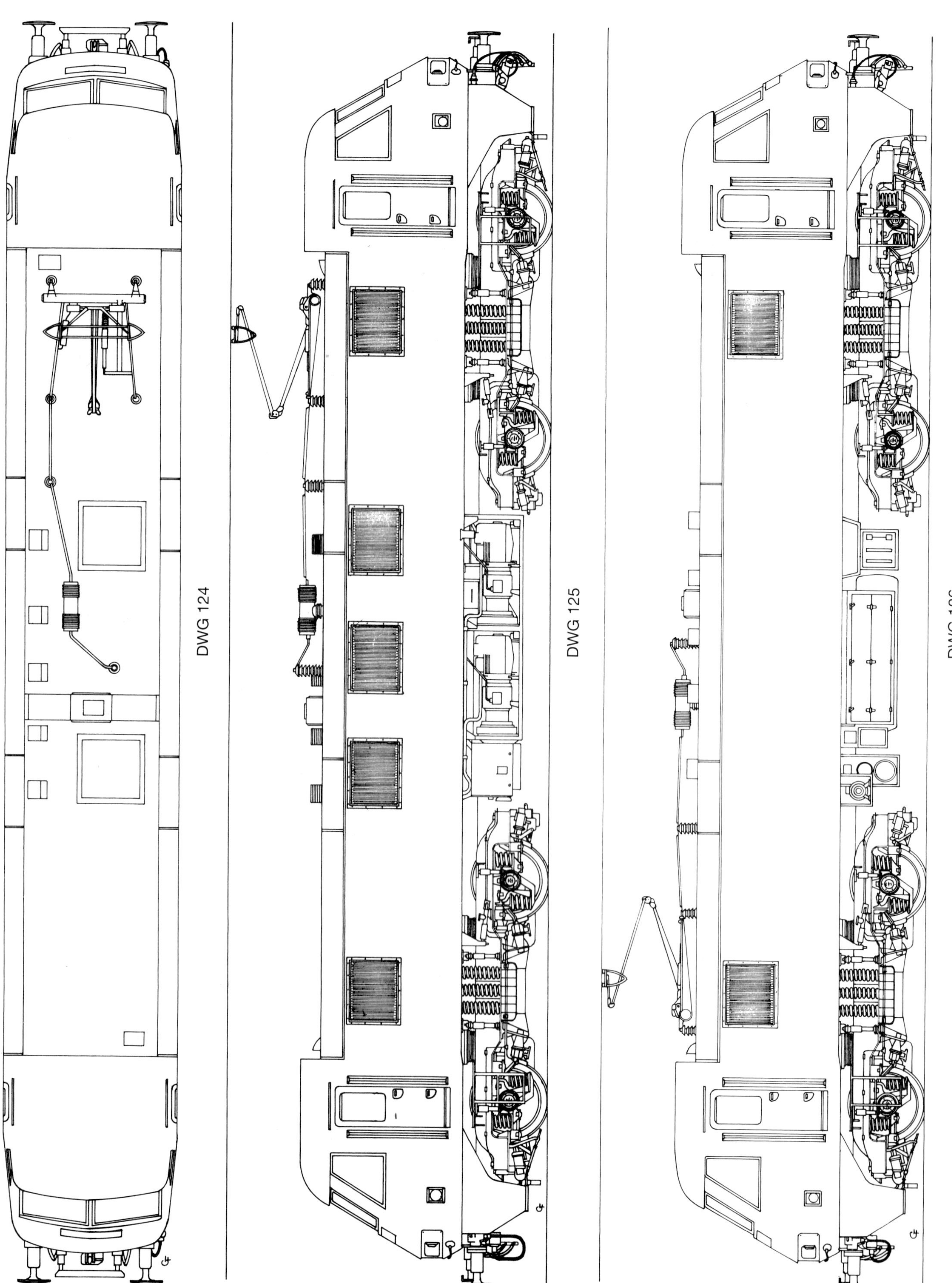

DWG 127

DWG 124
Class 90 roof detail. No. 1 end on right.

DWG 125
Class 90 side 'B' elevation. No. 1 end on right.

DWG 126
Class 90 side 'A' elevation. No. 1 end on left.

DWG 127
Class 90 front end detail fitted with buck-eye coupling and rubbing plate.

Class		90
Number Range	90/0	90001-90025
	90/1	90126-90150 (Note: 2)
Built by		BREL Crewe
Introduced		1987-88
Wheel Arrangement		Bo-Bo
Weight		85 tons
Height - pan down		13ft 0¼in
Width		9ft
Length		61ft 6in
Min Curve negotiable		4 chains
Maximum Speed		110mph (Note: 1)
Wheelbase		43ft 6in
Bogie Wheelbase		10ft 9in
Bogie Pivot Centres		32ft 9in
Wheel Diameter		3ft 9½ in
Brake Type		Air (Rheostatic)
Sanding Equipment		Pneumatic
Heating Type		Electric - Index 95 (Note: 2)

Route Availability		7
Coupling Restriction		TDM fitted
Brake Force		40 tons
Horsepower	(continuous)	5,000hp
	(maximum)	7,860hp
Tractive Effort	(maximum)	43,150lb
Number of Traction Motors		4
Traction Motor Type		GEC 412 BZ
Control System		Thyristor
Gear Ratio		32:73
Pantograph Type		Brecknell Willis HS
Nominal Supply Voltage		25kV ac

Note: 1 Class 90/1 maximium speed 75mph.

Note: 2 Class 90/1 locomotives are operated by the Railfreight business and are not fitted with buck-eye couplings or rubbing plates and have had their ETS equipment isolated.

By the mid-1980s BR were seeking new generation electric locomotives for its WCML services, and government authorisation was sought for a fleet of 50 thyristor controlled state-of-the-art locomotives during 1984. Permission for the build was eventually granted, with construction being awarded to BREL Crewe Works, and the main equipment sub-contractor being GEC Traction. The locomotives were to be classified 87/2.

The fleet were a break from previous 25kV ac designs in having steeply raked back cab ends, and for the first time on a production locomotive type, were fitted with Time Division Multiplex (TDM) multiple control equipment.

Construction at BREL Crewe commenced in late 1986, with the first locomotive being shown off to the public at a July 1987 open-day. The first locomotive, numbered 90001, was however not finished until September 1987. During the course of construction the classification was changed from Class 87/2, to 90 as the new product showed virtually no physical or technical resemblance to the Class 87. After completion at Crewe the first locomotive was transferred to the Railway Technical Centre at Derby for a major 'type test' programme.

Delivery of the 50 locomotives was a protracted affair, with the final locomotive not being handed over to the operators until the end of 1990. The Class 90s were designed for mixed traffic (passenger and freight) operation. The batch allocated to InterCity were first off the production line, and once tests had been completed and sufficient stock fitted with TDM introduced, together with Driving Van Trailers being commissioned, the locomotives commenced operation on Euston – Birmingham, North West and Anglo-Scottish duties, working in the push-pull mode. Subsequent Class 90s were constructed for the Railfreight Distribution (RfD) business for use on high speed long-distance block freight services. By 1992 the allocation of the class was again amended with a handful operating for Rail express systems (Res), the railway parcels arm.

A major feature of the Class 90 was the installation of drop-head buck-eye couplers with retractable side buffers – a feature essential for high-speed push-pull operation. Although all locomotives were built with this fitting, those operated by RfD have since had it removed and ETS equipment isolated.

The livery applied to the first 25 Class 90s, when constructed, was full InterCity, with the next eleven emerging in Main Line livery and the final 14 were painted in Railfreight Distribution colours. Following the introduction of the Res red livery the five locomotives operated by this business have been repainted into the house colours.

After the splitting of the railway businesses most of the Class 90s funded by Railfreight were reclassified Class 90/1, to identify the isolation of their ETS equipment, renumbering was made in the 901xx series.

The InterCity Class 90s are allocated to Willesden depot, with the remainder, (Nos 90021-025/126-150) being shedded at Crewe. During 1990, when the ECML became fully electrified, the Class 90s started to appear on some main line duties from King's Cross, mainly deputising for non-available Class 91s. From the commencement of the winter timetable in 1991 the class were rostered for some ECML services, and from the commencement of the Summer 1992 timetable, the Res owned locomotives have had rostered van workings over the full length of the ECML.

The construction contract for the Class 90 build was awarded to BREL Crewe, with the chief sub-contractor being GEC Traction. The first complete Class 90 was finished at Crewe during October 1987, with the various assembly shops effecting the remainder of the build over the next two years. This view of the main shop taken on 12th February 1988 shows locomotive No. 90005 undergoing static tests prior to operating on the main line for the first time.

Colin J. Marsden

Prior to the first Class 90 emerging from Crewe, the classification given to the Class was 87/2, however as these locomotives bore little physical or technical resemblance to the Class 87 fleet it was decided to allocate a different class designation. Prototype locomotive No. 90001 is seen in the advance stages of construction in Summer 1987.

Colin J. Marsden

The eighth member of the Class 90 build, No. 90008, takes shape in the new build shop at Crewe on 12th February 1988. At this stage the steel body sides and reinforced glass fibre cabs are all in situ, but the technical assembly has yet to be undertaken.

Colin J. Marsden

After completion at BREL Crewe, No. 90001 was transferred to the Engineering Development Unit (EDU) of the Railway Technical Centre, Derby where virtually every item of equipment was placed under a type test procedure. No. 90001 is seen inside the EDU whilst being prepared for an active test programme.

Colin J. Marsden

Throughout early 1988 the Class 90 locomotives were engaged in a major driver training programme involving several hundred staff. One of the driver training runs for Crewe based drivers is seen departing from Crewe on 16th February 1988 behind No. 90003 bound for Carlisle.

John Tuffs

On 29th March 1990, InterCity owned No. 90003 passes near the village of Clifton, south of Penrith with the 10.25 Inverness – Euston service, which the Class 90 would have worked forward from Glasgow.

Colin J. Marsden

Again carrying InterCity livery, No. 90024, which is now owned by Railfreight, but in 1992 was still maintained to main line passenger standards hurries down Hest Bank on 28th March 1990 with the 15.15 Edinburgh – Paddington service. The class 90 would be replaced by a Class 47 diesel forward from Birmingham New Street.

Colin J. Marsden

In far from ideal photographic conditions No. 90006 races through Warrington Bank Quay station on 6th October 1988 with a southbound driver training special.

Brian Morrison

Following the introduction of the Rail express system trading title from October 1991, the new business identity of red and grey, off-set by light blue bands was applied to some traction owned by the business. The first Class 90 to carry the colours was No. 90020, which was named *Colonel Bill Cockburn CBE TD*, and is shown here at Crewe.

Colin J. Marsden

On 30th March 1990, main line liveried Class 90 No. 90033, now renumbered to 90133 starts the climb of Hest Bank with the 07.20 Penzance – Glasgow/ Edinburgh service. This locomotive is now painted in RfD colours.

Colin J. Marsden

For a short period in Spring 1990, Class 90s could be found at the head of coal trains on the East Anglia main line, while driver training on the traction between Ilford and Ipswich was being carried out. On 5th April 1990, No. 90028 hurries towards Colchester with the 14.00 Ipswich – Ilford training special.

Colin J. Marsden

With the last three digits of its number 145 on the front end, Class 90/1 No. 90145 in RfD livery passes Plumpton, north of Penrith on 17th September 1992, with a daily Crewe Basford Hall – Mossend freight.

Colin J. Marsden

Class 91

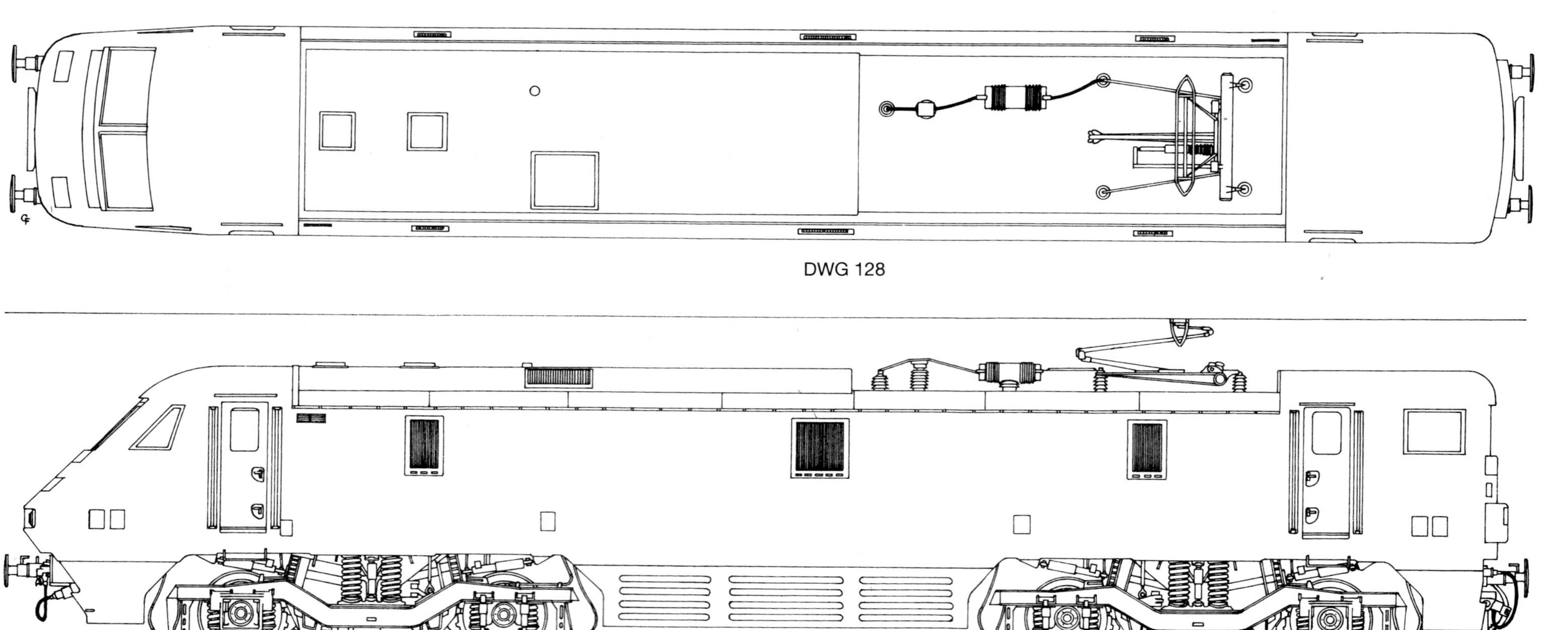

DWG 128

DWG 129

DWG 130

DWG 128
Class 91 roof detail. No. 1 end on left.

DWG 129
Class 91 side elevation. No. 1 end on left.

DWG 130
Class 91 front end detail of No. 1 (raked) end.

DWG 131
Class 91 front end detail of No. 2 (slab) end.

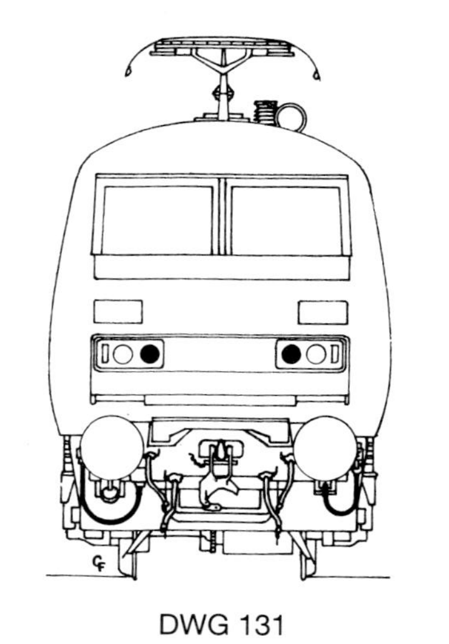

DWG 131

Class	91
Number	91001-91031
Built by	BREL Crewe & GEC-TPL
Introduced	1988-91
Wheel Arrangement	Bo-Bo
Weight	80 tons
Height – pan down	12ft 4in
Width	9ft
Length	63ft 8in
Min Curve negotiable	4 Chains
Maximum Speed	140mph*
Wheelbase	45ft 4½in
Bogie Wheelbase	10ft 11⅞in
Bogie Pivot Centres	34ft 5½in
Wheel Diameter	3ft 3½in
Brake Type	Air (Rheostatic)
Sanding Equipment	Pneumatic
Heating Type	Electric - Index 95
Route Availability	7
Coupling Restriction	TDM fitted
Brake Force	45 tons
Horsepower (continuous)	6,090hp
(maximum)	6,300hp
Traction Motor Type	GEC G426AZ
Control System	Thyristor
Gear Ratio	1 74:1
Pantograph Type	Brecknell Willis HS
Nominal Supply Voltage	25kV ac

* Restricted to 125mph running until full ATP introduced.

In 1984, after the APT project had almost petered out, the design and indeed implementation of new locomotives and rolling stock for the WCML were of prime importance to the business directors. Thus the InterCity 225 project was born; this calling for the construction of APT-P style power cars (with cabs), a rake of trailer vehicles and a fleet of Driving Van Trailers (DVTs). Potential builders were invited to prequalify for approximately 25 sets in Autumn 1984. Amongst strict guidelines was the requirement that the locomotive must have the ability to haul express passenger services by day and lower speed sleeper or Freightliner duties by night.

Also towards the autumn of 1984 came the authorisation for East Coast Main Line (ECML) electrification. Originally the Class 89 Co-Co design was to be perpetuated for this route, but soon, the InterCity team decided to opt for the IC225 principle on this route, and by early 1985 the IC225 team was formed as an ECML project. By April 1985, under the competitive tendering policy, three firms were invited to submit building tenders – ASEA of Sweden, Brush and GEC Transportation projects (GEC-TPL). After much deliberation the contract was placed with GEC in February 1986. GEC-TPL were given the complete design, build and test brief, although GEC-TPL sub-contracted mechanical construction to BREL Ltd at Crewe. Under BR's numerical classification the locomotives for this project became Class 91.

Following signing of the contract in February 1986 it was announced that the first locomotive would be rolled out on 14th February 1988 – in just two years. After this announcement the design and construction went full speed ahead, despite many problems being encountered before the St Valentine Day roll out. Many new and novel design features were incorporated in the BR Class 91 build, including the provision of a streamlined or raked back, No. 1 end for high speed operation, and a slab fronted No. 2 end for slower operation. One of the most significant changes on this type of locomotive was in the bogie and traction equipment design, incorporating frame mounted traction motors driving an axle-mounted gearbox via a carden shaft. The disc brake unit is off wheel and fitted on the rear of the traction motor. The Class 91 also incorporates the very latest state-of-the-art computer-based electronics for power and brake control.

In common with all new high speed types, buck-eye couplers are fitted, as well as Time Division Multiplex (TDM) jumpers for remote control from DVT stock.

After the GEC-TPL/BREL Ltd roll out on 14th February 1988, No. 91001 commenced a series of tests, first at Crewe and then at the Railway Technical Centre, Derby. By late March No. 91001 was delivered to Bounds Green depot on the ECML where test running and training commenced. Due to the protracted deliveries of the purpose-built DVT vehicles, several IC125 (HST) power cars were adapted for TDM operation, and a limited passenger service headed by Class 91s commenced from October 1989. Full implementation of Class 91s did not commence until 1991, and even during 1992, some services scheduled for Class 91 operation were still being operated by IC125 stock.

The initial build of Class 91s was for just ten locomotives, which were introduced for evaluation running. After many thousands of hours of running the manufacturers and BRB then commenced construction of the remaining 21 locomotives to make a fleet of 31 in total. The final locomotive was completed at Crewe during February 1991.

The livery of the Class 91 is standard InterCity, offset by the usual yellow panel warning ends. Following their introduction many examples of the Class 91s have been named, including one locomotive No. 91029, *Queen Elizabeth II*.

By 1993 the IC225 or Class 91 train sets have settled down quite well, operating on the King's Cross – Leeds, York – Newcastle – Edinburgh and Glasgow route, operating at speeds up to 125mph. The full design speed of 140mph will not be possible until full automatic train protection equipment is designed and installed.

Although designed for working slab-end-first on non-passenger duties, this form of operation has not taken place, and is unlikely to be done in the foreseeable future.

BRE
BRITISH RAIL ENGINEERING LIMITED
CREWE WORKS
BRE
BRITISH RAIL ENGINEERING LIMITED
CREWE WORKS
INTERCITY
GEC
GEC

Left: Although constructed at BREL Ltd Crewe works, the Class 91 construction contract was awarded to GEC, who sub-contracted BREL to build the mechanical portion. In September 1990 the body shell of No. 91025 is seen almost complete in the main erecting shop.

Colin J. Marsden

Left, below: The first Class 91 was rolled out of BREL Crewe for the press to inspect on 12th February 1988, two days before the locomotive was handed over to BR for trial running. Looking immaculate No. 91001 stands in the works yard, viewed from its No. 1 end.

Colin J. Marsden

Running slab end forward, No. 91002 passes near Sandy during August 1988 with a southbound test special from Peterborough to Bounds Green. The stock in this train is of the Mk 1 and Mk 2 types.

Brian Morrison

Once sufficient Class 91 locomotives were available a major training and testing operation commenced on the ECML, with a 24 hour per day programme being operated. On 10th August 1988 No. 91004 storms past Sandy with the 12.38 King's Cross – Doncaster training special, the train being formed of test car *Prometheus*, five Mk 3 sleeping cars and a DVT.

Brian Morrison

Much of the staff training on the Class 91s was undertaken at Doncaster. During the evening of 10th March 1989, Nos 91003 and 91006 stand in Doncaster station yard, coupled to a driver training special. The following morning the two locomotives were uncoupled, and operated on their own.

Colin J. Marsden

With a rake of IC125 stock behind, Class 91 No. 91008 hurries through Doncaster on the middle road on 25th April 1989, with the 07.50 King's Cross – Leeds "Yorkshire Pullman" service. This was one of the first ECML services to be taken over by Class 91 traction.

Colin J. Marsden

With a rake of IC125 passenger stock behind, No. 91008 pulls out of Wakefield Westgate on 25th April 1989 with the 14.20 King's Cross – Leeds. At this time, as no purpose-built Mk 4 DVT vehicles were available, modified IC125 power cars were coupled at the London end of sets.

Colin J. Marsden

Leading a full rake of GEC-Alsthom built Mk4 stock, Class 91 No. 91026 pulls towards York through Holgate Bridge Junction on 13th July 1992, while forming the 08.00 King's Cross – Glasgow service.

Colin J. Marsden

Arriving at the remodelled York station, rebuilt as part of the ECML Total Route Modernisation package, Class 91 No. 91016 powers the 11.30 King's Cross – Glasgow service on 13th July 1992. From the commencement of the Summer 1992 timetable, the fastest services between King's Cross and Edinburgh were timed for just 3hr 59min.

Colin J. Marsden

Propelling a formation of Mk4 stock, Class 91 No. 91018 departs from Glasgow Central on 21st August 1992 with the 18.00 to King's Cross. With the Class 91 coupled at the rear, the driver controls the locomotive from the DVT via the Time Division Multiplex system.

Colin J. Marsden

Detail of Class 91 driving cab

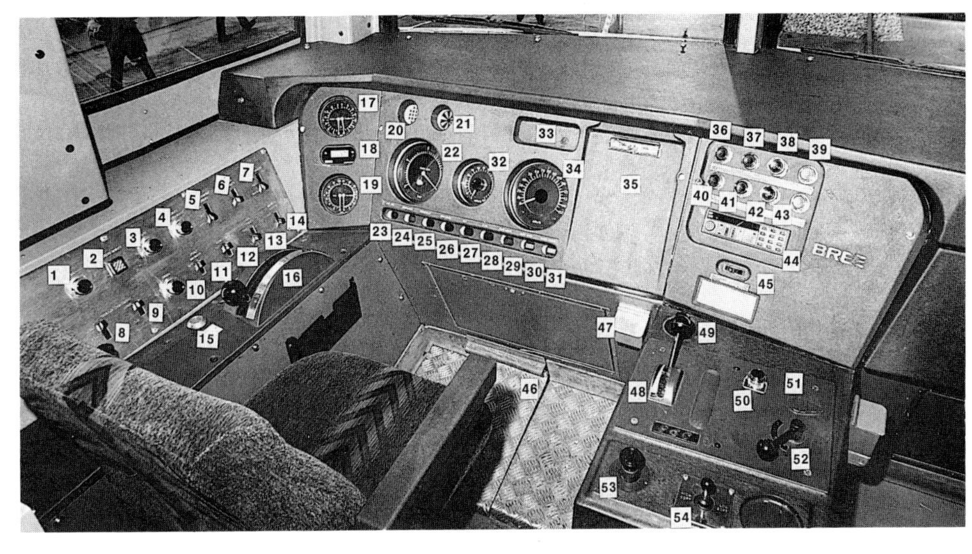

1. Park brake on button
2. Park brake indicator
3. Park brake off button
4. Brake overcharge switch
5. Instrument light switch
6. Cab light switch
7. Clip board light switch
8. Cab air treatment switch, heat/vent/cool
9. Cab air treatment switch, high/low
10. Sand button
11. Foot warmer switch
12. De-mister switch
13. Tail light switch
14. Marker light switch
15. Emergency brake plunger
16. Brake controller
17. Bogie brake cylinder gauge
18. Clock
19. Main reservoir gauge
20. AWS alarm
21. AWS indicator
22. Brake pipe gauge
23. AWS isolate warning light
24. Headlight warning light
25. Wheelslip warning light
26. General fault light
27. Tilt warning light
28. Electric train supply (ETS) warning light
29. Line light
30. Pantograph auto-drop light
31. High Speed brake indicator
32. Pre-set speed control
33. Space for ATP
34. Speedometer
35. Noticeboard
36. Driver/guard call button
37. ETS on button
38. Passenger communication override button
39. Pantograph up or re-set button
40. Fire alarm test button
41. ETS off button
42. Fire extinguisher delay button
43. Pantograph down button
44. Loco/shore radio system
45. Headlight switch
46. Drivers' Safety device foot pedal
47. Ashtray
48. Power controller
49. Windscreen wiper control
50. Tractive effort boost button
51. Master switch key socket
52. Master switch
53. AWS re-set button
54. Horn valve

One of the most important days for the electrification of the ECML was on 28th June 1991, when Her Majesty The Queen visited the King's Cross – Edinburgh route, travelled on a Class 91 hauled Mk4 set and named Class 91 No. 91029 *Queen Elizabeth II* prior to departure from King's Cross. On the same day the Queen also named another Class 91, *Palace of Holyroodhouse.* The picture below left shows the Queen unveiling the plate on the side of No. 91029, while the view right, is of the Queen departing from the cab of the locomotive after inspecting the controls, with Sir Bob Reid, the BR Chairman.

Both: Colin J. Marsden

Towards the future

The development of electric traction for Britain's railways is continuing all the time; as this book closes for press in early 1993, BR are still awaiting the delivery of the first Class 92 dual-powered electric locomotive. The fleet of 37 25kV ac/750V dc locomotives will be dedicated to Channel Tunnel use, and operate freight/passenger trains through the Tunnel to a terminal just inside France. In England the locomotives will be allocated to Crewe and operated by Railfreight Distribution and European Passenger Services.

Although numerous drawings and basic line drawings have been produced on the Class 92, the fleet has not been included in the main stream of this book as the final drawings were not available from manufacturers, Brush at the beginning of 1993.

Other new electric locomotive traction that will be seen in this country will be the Bo-Bo-Bo Eurotunnel locomotives for operation on the Channel Tunnel shuttle services. These locomotives are again being built by Brush, with the first locomotive delivered in early 1993.

The long term future of electric locomotive operation in this country is good, although at present a lack of investment in Britain's railways precludes the ordering of WCML replacement traction, but when this is available a fixed train consist system, similar to the IC225s on the East Coast Main Line are the most likely to be ordered.

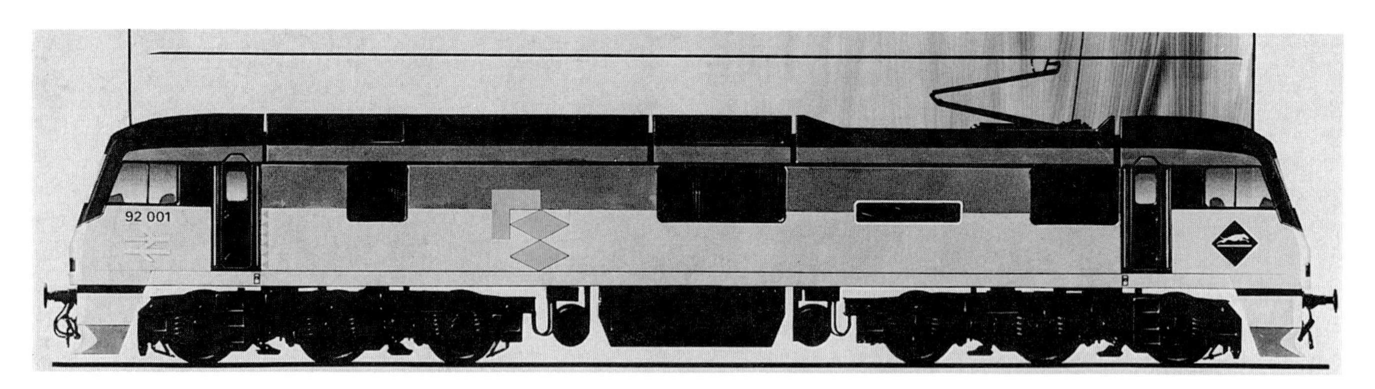

Latest drawing of the Class 92. This illustration shows the locomotive in RfD colours, but when the first locomotive is delivered it should be painted in the new International livery, which is more representative of the passenger/freight operations the traction will work. *ABB*

At last year's Freightconnection exhibition in Birmingham, a full size mock-up of a Class 92 was the centre piece of the entrance display. This showed for the first time the style of the livery and the front-end technical layout, with a high intensity headlight on the roof line, and control jumper cables on the buffer beam, provided to control a locomotive at each end of trains passing through the tunnel. *Colin J. Marsden*